Reshaping College Mathematics

A project of the
Committee on the Undergraduate Program in
Mathematics

Edited by
Lynn Arthur Steen

MAA Notes Series

The MAA Notes Series, started in 1982, addresses a broad range of topics and themes of interest to all who are involved with undergraduate mathematics. The volumes in this series are readable, informative, and useful, and help the mathematical community keep up with developments of importance to mathematics.

First Printing
© 1989 by the Mathematical Association of America
Library of Congress Number 89-061338
ISBN-0-88385-062-1
Printed in the United States of America

Reshaping College Mathematics

Reshaping College Mathematics

Preface

Calls for Change

For over 35 years the Committee on the Undergraduate Program in Mathematics (CUPM) has helped provide coherence to the mathematics major by monitoring practice, advocating goals, and suggesting model curricula. This volume brings together various curriculum reports issued during the decade of the 1980's. It provides a convenient reference for the mathematical community as it begins to reshape college mathematics in response to mounting demands for change.

Many of the calls for reform have been expressed in published reports, for example:

1989 *Curriculum and Evaluation Standards for School Mathematics*, published by the National Council of Teachers of Mathematics.

1989 *Everybody Counts: A Report to the Nation on the Future of Mathematics Education*, published by the National Research Council.

1988 *Changing America: The New Face of Science and Engineering*, published by the White House Task Force on Women, Minorities, and the Handicapped in Science and Technology.

1987 *The Underachieving Curriculum*, published by the Second International Mathematics Study.

1986 *Towards a Lean and Lively Calculus*, published by the Mathematical Association of America.

1985 *Integrity in the College Curriculum: A Report to the Academic Community*, published by the Association of American Colleges.

1984 *Renewing U.S. Mathematics: Critical Resource for the Future*, published by the National Research Council.

Other pressures for change are expressed in articles diffused throughout the literature on a wide range of issues, from remediation (too much) through Ph.D. production (too little), from students (greater diversity) to mathematics (greater applicability), and from technology (under-utilized) to pedagogy (too passive). What all reports have in common is the case they make for significant change in undergraduate mathematics to serve better the needs of students who will live and work in the twenty-first century.

Background

CUPM was established in 1953 to "modernize and upgrade" the mathematics curriculum and to halt what was even then decried as "the pessimistic retreat to remedial mathematics." At that time total enrollment in college mathematics courses in the United States was approximately 800,000; each year about 4,000 students received a bachelor's degree in mathematics, and about 200 received Ph.D. degrees.

In its early years CUPM concentrated on proposals to strengthen undergraduate preparation for graduate study in mathematics. Spurred on by Sputnik and assisted by significant support from the National Science Foundation, the mathematical community in the United States matured rapidly from a servant discipline indentured to science and engineering to vigorous world leadership. By 1970 U.S. mathematics departments produced 24,000 bachelor's and 1,200 doctoral degrees—a six-fold increase in less than twenty years.

Then the bubble burst. As student interest shifted from personal goals to financial security, and as computer science began to attract increasing numbers of students who in earlier years might have studied mathematics, the numbers of mathematics bachelor's degrees dropped by over 50% in ten years, as did the number of U.S. students who went on to a Ph.D. in mathematics. In 1981, at the nadir of B.A. productivity, CUPM published a comprehensive report entitled *Recommendations for a General Mathematical Sciences Program*. Significantly, this report advocated not a strengthened program in traditional (pre-doctoral) mathematics, but a broad, innovative program in *mathematical sciences*.

Although CUPM did not create the movement towards mathematical sciences, its 1981 report helped legitimize a process that was well under way. As a consequence, mathematics programs in U.S. colleges and universities are now dominated by variations on two paradigms that reflect the two phases of CUPM activity. The first, a fading image of the CUPM recommendations of the 1960's, focuses on core mathematics as preparation for graduate study in mathematics. The second, reflecting the broader objectives of CUPM's 1981 mathematical sciences report, focuses on mathematical tools needed for a "life-long series of different jobs." Most campuses support a mixed model representing a locally devised compromise between these two standards.

Issues of the 1980's

During the 1980's several issues emerged that had great bearing on the conduct of undergraduate mathematics. Pressure from the computer science community created a demand for a freshman or sophomore course in discrete mathematics. This posed issues of definition (what was to be included?), level (how much maturity was required?), and articulation (where could it fit into the ubiquitous calculus sequence?). No single answer emerged, and experiments continue to determine locally optimal strategies for meeting this important new need.

As American society moved towards greater concern with material well-being, pressure from many sources—not the least being from parents and schools in affluent districts—created enormous demand for high school calculus. Suddenly large numbers of students came to college with uneven preparation that partially overlapped standard introductory college courses. Problems of articulation between high school leaving and college entering became quite intense. The new *Standards* for school mathematics of NCTM promise to increase the diversity of student preparation in years ahead, as some districts adopt new programs and others retain old habits.

From a different source—mostly from parents and public officials concerned with the quality of higher education—came calls for assessment and evaluation to ensure that all students receive certain minimal skills from their college study. Quantitative literacy (or "numeracy") joined the litany of demands generated by discussions of "cultural literacy" and "competitiveness." Quantitative competence and mathematical appreciation of students who do not study mathematics for professional reasons became—and still is—a major concern on college campuses.

The Mathematical Association of America responded to each of these issues—discrete mathematics, school articulation, quantitative literacy—by a variety of studies, some under CUPM, some in cooperation with other organizations. Issued at different times throughout the past decade, these studies supplement CUPM's comprehensive recommendations for an undergraduate program with specific recommendations in areas of timely concern. They are all gathered in this volume, where together they provide a thorough airing of issues pertinent to reshaping college mathematics.

The More Things Change ...

Despite the many new issues that have arisen in recent years (e.g., desk-top workstations, changing demographics, calculus reform), curriculum reports in this volume—which were generally written three to ten years ago—align remarkably well with contemporary calls for change that one hears at every professional meeting. Today's advice, by and large, is no different than yesterday's. It is just being said with greater urgency.

Here, for example, is a sample of recommendations to be found among the reports in this volume:

On Goals:

- A mathematical sciences major should develop a student's capacity to undertake intellectually demanding mathematical tasks.
- A major in mathematical sciences should emphasize general mathematical reasoning as much as mastery of various subject matter.
- The instructor's central goal should be to teach students how to learn mathematics, expecting that students will correctly retain only a tiny portion of what was taught.
- A mathematical sciences curriculum should be designed around the abilities and academic needs of the average student, with supplementary work available to attract and challenge more advanced students.
- College students must understand the historical and contemporary role of mathematics and be able to place the discipline properly in the context of other human intellectual achievement.
- Students should gain an ability to read and learn mathematics on their own. Such maturity is as much a function of *how* mathematics is learned as *what* is learned.

On Teaching:

- Students should be led to discover mathematics for themselves, rather than merely being presented with the results of concise, polished theories.
- The approach to most topics should involve an interplay of applications, problem-solving, and theory. Applications should motivate theory so that theory is seen by students as useful and enlightening.
- Freshman courses in mathematics should be designed to appeal to as broad an audience as is academically reasonable.
- In the first two years, theorems should be *used* rather than *proved*. The place for theoretical rigor is in later upper-level courses.
- The greatest challenge is that students enter college with much less mathematics than they used to, but they expect to leave with much more.

ON COMPUTING:

- Students should make full use of calculators and computers in all mathematics courses.
- All mathematical sciences students must be given an introduction to the basic concepts of computer science.
- Computing assignments should be used in most mathematics courses.

ON MODELING:

- Applications and modeling should be included in a nontrivial way in most college-level mathematical sciences courses.
- All students majoring in mathematics should undertake some real-world mathematical modelling project.

ON WRITING:

- Explaining a mathematical result in terms of a real-world setting involves the need to communicate in a precise and lucid manner. This aspect of a mathematical scientist's training should not be left to courses in other sciences or to on-the-job learning after graduation.
- Teachers of mathematics should employ strategies that encourage student reading, writing, and reflection.
- Students should be asked to make formal oral and written presentations. A (non-original) paper serves the dual purpose of developing communication skills and introducing pedagogical flexibility.

ON STATISTICS:

- New knowledge has rendered a course devoted solely to the theory of classical parametric procedures out of date.
- The traditional undergraduate course in statistical theory has little contact with statistics as it is practiced and is not a suitable introduction to the subject.

ON DISCRETE MATHEMATICS:

- Discrete mathematics at the intellectual level of calculus should be part of the standard mathematics curriculum in the first two years.
- Topics covered are less important than acquiring mathematical maturity and skills in using abstraction and generalization.
- Mathematics majors should be required to take at least one course in discrete mathematics.

ON CALCULUS:

- Students should learn the content of the full four years of high school mathematics before enrolling in calculus.
- Calculus in high school should be taught with the expectation that successful graduates would not repeat calculus in college.
- Colleges need to provide individualized placement for students who have studied calculus in high school.

ON MINIMAL EXPECTATIONS:

- All college graduates should be expected to demonstrate reasonable proficiency in the mathematical sciences.
- College remedial courses should not be a rehash—and certainly not an accelerated rehash—of traditional school courses. Students should find even remedial courses fresh, interesting, and significant.

A Context for Reform

One must wonder, after reviewing all the arguments produced by the various committees whose reports are contained in this volume, why so little has changed. Why did it take seven years from the time that CUPM urged in 1981 that all beginning courses must be taught in an effective and attractive manner for the community to take a hard look at calculus? Why is it only now rather than in 1981 that mathematicians are beginning to realize the importance of writing assignments—both to learn to write and to write to learn?

Momentum may be one reason. In the early 1980's there was very little support for educational change. The political agenda of the nation at that time was not supportive of issues in science and mathematics education. All educational activities at the National Science Foundation were eliminated in 1980, only to be restored several years later. MAA and NCTM released separate reports (*Prime 80, An Agenda for Action*) into this thin atmosphere of uncertainty about science and mathematics education. It should not be surprising to find, ten years later, that most of the problems identified in these reports are still evident today.

Today, in contrast to 1980, many different organizations are working together for the improvement of mathematics education. The National Academy of Sciences, the National Science Foundation, and many private foundations have joined with the several mathematical societies to work on a common plan for revitalizing mathematics education. Now, after a decade of talking, everyone is finally moving in the same direction.

A partial list of current activities that relate to college mathematics reveals clearly the breadth of current support for reshaping college mathematics:

- The Undergraduate Curriculum Initiative at the National Science Foundation featured calculus reform in its first wave of proposal solicitations.

- The Mathematical Sciences Education Board and the Board on Mathematical Sciences at the National Research Council have jointly established the Committee on Mathematical Sciences in the Year 2000 to analyze collegiate mathematics and make recommendations for improvement.

- The Division of Mathematical Sciences at the National Science Foundation is now supporting research experiences for undergraduates, and is planning to add educational dimensions to many of its new initiatives.

- The MAA Committee on the Mathematical Education of Teachers is working with NCTM and with the National Board for Professional Teaching Standards to revise the recommendations for the mathematical education of teachers of mathematics in a manner consistent with the new NCTM *Standards*.

- Both AMS and SIAM now have committees dealing with education, as well as liaison members on CUPM.

- The MAA Committee on the Undergraduate Program in Mathematics has subcommittees working on recommendations for calculus and other courses in the first two years, on the mathematical sciences major, on service courses, and on the role of symbolic computer systems.

- MAA has recently published two volumes of papers dealing with calculus, one report on the role of computers in undergraduate mathematics, and one report on the continuing mathematical education of teachers. A report on discrete mathematics is forthcoming.

The present volume provides wisdom to support these efforts. It reflects the best thinking of many experienced mathematicians and teachers who have struggled with curricular questions facing college mathematics as part of their work for CUPM and other MAA committees. Although certain sections are obviously dated (e.g., discussions of computing, reference lists), the central message of this volume provides a philosophy of instruction that is as valuable now as when it was written.

We don't need to look far for sound goals and objectives for college mathematics. Most of what we need can be found in this volume. What remains to be done—as much now as ever—is to find effective means of turning ideals into practice.

Lynn Arthur Steen, Chair
Committee on the Undergraduate
   Program in Mathematics
St. Olaf College
March, 1989

Mathematical Sciences

In 1981 the Committee on the Undergraduate Program in Mathematics (CUPM) published a major report entitled RECOMMENDATIONS FOR A GENERAL MATHEMATICAL SCIENCES PROGRAM. *This report comprises six chapters that are reprinted here, with minor editing, as the first six chapters of the present volume. Alan Tucker, Chairman of the CUPM Panel that wrote the 1981 report, has written a new Preface to introduce this reprinting.*

1989 Preface

In the eight years since the CUPM Recommendations on a General Mathematical Science Program appeared, issues in mathematics curriculum, such as calculus reform and discrete mathematics, have become hot topics in the mathematics community and have even received extensive coverage in the popular press. The CUPM Panel on a General Mathematical Sciences Program had the luxury of working in comparative anonymity, although ten panel discussions at national and regional mathematics meetings gave the panel some professional visibility. The Panel's basic goal was to give long-term, general objectives for undergraduate training in mathematics.

The 1960's and 1970's had seen a variety of specialized appeals made to college students interested in mathematics. For example, the discipline of computer science emerged as an exciting career for mathematics students. The earliest CUPM recommendations for the mathematics major were aimed at preparing students for doctoral work in mathematics. By the late 1970's, there was a sense that the mathematics major had lost its way, with upper-division enrollments in traditional core courses like analysis and number theory down by 60% from their levels five years earlier and with industrial employers showing little interest in hiring mathematics majors.

To put these recent events in perspective, the Panel obtained a historical briefing from Bill Duren (the founding chairman of CUPM). He recounted over a century of swings of the pendulum between the theoretical and the practical in American collegiate mathematics education, and between training for careers of the future and training in classical, old-fashioned methods.

The Mathematical Sciences Panel sought to find a common ground for the mathematics major which taught abstraction and application, emerging new problem areas and time-tested old ones. The Panel sought to persuade mathematicians that the curriculum in the mathematics major should be shared among the various intellectual and societal constituencies of mathematics. The challenge was to be diverse without being superficial.

The most concrete consequence of the Panel's work was its name, Panel on a General Mathematical Sciences Program. It asked that the mathematics major be renamed the mathematical sciences major—a change explicitly adopted by hundreds of colleges and universities and implicitly adopted by the vast majority of institutions. The Panel recommended that first courses in most subjects should have a good dose of motivating applications, particularly linear algebra and statistics, and that one advanced course should have a mathematical modeling project. This recommendation also seems to have wide acceptance. There were several panel recommendations that reflected trends already occurring but being resisted by some mathematicians: requiring an introductory course in computer science; not requiring linear algebra as a prerequisite for multivariable calculus; encouraging weaker students to delay core abstract courses until the senior year; and not requiring every mathematics major to take courses in real analysis and abstract algebra (i.e., other mathematics courses at comparable levels of abstraction could be substituted).

Although it was unhappy with calculus, the Mathematical Sciences Panel consciously avoided recommending changes in calculus for fear that the inevitable controversy and the complexity of such an undertaking would undermine acceptance of its basic recommendations about the structure of a mathematics major. The Panel touched only lightly on the issue of discrete versus continuous mathematics, recommending exposure to "more combinatorially-oriented mathematics associated with computer and decision sciences" (Tony Ralston's provocative essays about discrete mathematics had not yet appeared).

It was gratifying to the Mathematical Sciences Panel that its report was well-accepted: all two-thousand copies printed have been sold (another two-thousand copies had been sent gratis to department heads). In reviewing the report for this reprinting, the only changes have been to add a few additional references. On the other hand, there was one panel suggestion that has been ignored thus far and which merits consideration.

It concerns the "modest" version of abstract algebra (in Section III) in which time would be spent sensitizing students to recognize how algebraic systems arise naturally in many situations in other areas of mathematics and outside mathematics (to keep algebra alive in their minds after they leave college).

ALAN TUCKER
SUNY at Stony Brook
March, 1989

1981 Preface

This report of the CUPM Panel on a General Mathematical Sciences Program (MSP) presents recommendations for a mathematical sciences major. The panel has concentrated its efforts on general curricular themes and guiding pedagogical principles for a mathematical sciences major. It has tried to frame its recommendations in general terms that will permit a variety of implementations, tailored to the needs of individual institutions. A prime objective of the original 1960's CUPM curriculum recommendations for upper-level mathematics courses was easing the trauma of a student's first year of graduate study in mathematics. This report refocuses the upper-level courses on the traditional objectives of general training in mathematical reasoning and mastery of mathematical tools needed for a life-long series of different jobs and continuing education.

The MSP panel has tried to avoid highly innovative approaches to the mathematics curriculum. The emphasis, instead, has been on using historically rooted principles to organize and unify the mathematical sciences curriculum. The MSP panel believes that the primary goal of a mathematical sciences major should be to develop rigorous mathematical reasoning. The word 'rigorous' is used here in the sense of 'intellectually demanding' and 'in-depth.' Such reasoning is taught through a combination of problem solving and abstract theory. Most topics should initially be developed with a problem-solving approach. When theory is introduced, it usually should be theory for a purpose, theory to simplify, unify, and explain questions of interest to the students.

CUPM now believes that the undergraduate major offered by a mathematics department at most American colleges and universities should be called a Mathematical Sciences major. Enrollment data show that for several years less than half the courses, after calculus, in a typical mathematics major have been in pure mathematics. Furthermore, applied mathematics, probability and statistics, computer science, and operations re-

search are important subjects which should be incorporated in undergraduate training in the general area of mathematics.

Computer science has become such a large, multifaceted field, with ties to engineering and decision sciences, that it no longer can be categorized as a mathematical science (at the National Science Foundation, computer science and mathematical sciences are different research categories). A mathematical sciences major must involve coursework in computer science because of the usefulness of computing and because of computer science's close ties to mathematics. Undergraduate majors in mathematical sciences and in computer science should complement each other.

The new course recommendations presented in this report do not, in most instances, replace past CUPM syllabi. They describe different approaches to courses; for example, a one-semester combined probability and statistics course, or a multivariate calculus course without a linear algebra prerequisite.

The work of the CUPM Panel on a General Mathematical Sciences Program was supported by a grant from the Sloan Foundation. The chairmen of CUPM during this project, Donald Bushaw and William Lucas, deserve special thanks for their assistance.

For information about other CUPM documents and related MAA mathematics education publications, write to: Director of Publications, The Mathematical Association of America, 1529 Eighteenth Street, N.W., Washington, D.C. 20036.

ALAN TUCKER
SUNY at Stony Brook

Panel Background

The CUPM Panel on a General Mathematical Sciences Program (MSP) was constituted in June, 1977 at a CUPM conference in Berkeley. CUPM members decided that a major re-examination of the mathematics major was needed. The CUPM model for the mathematics major contained in the 1965 CUPM reports on Pregraduate Training in Mathematics and a General Curriculum in Mathematics in Colleges (revised in 1972) was felt to be out of date. Following a six-month study, MSP reported to CUPM that the CUPM mathematics major curriculum should be substantially revised and broadened to define a mathematical sciences major. MSP was charged then with developing mathematical sciences recommendations.

Five subpanels were created to develop course recommendations in:

- The calculus sequence,
- Computer science,
- Modeling and operations research,
- Statistics, and
- Upper-level core mathematics.

The MSP project has had the cooperation of curriculum groups in the American Statistical Association, the Association for Computing Machinery, the Operations Research Society of America, and the Society for Industrial and Applied Mathematics. Graduate programs in the subjects covered by those societies draw heavily on undergraduate mathematics students, and except for computer science, undergraduate courses in these subjects are usually taught by mathematicians. Hence these curriculum groups had a major interest in the design of a mathematical sciences major.

The MSP panel coordinated its work with the National Research Council's Panel on Training in Applied Mathematics (chaired by P. Hilton, a member of MSP). The Hilton panel had a much broader mandate than the MSP panel. Its report addresses the unification of the mathematical sciences, the attitudes of mathematicians, academic-industrial linkages, and society's image of the mathematical sciences, as well as curricula. The Hilton report presented a limited number of general curriculum principles with the expectation that the MSP panel would develop fuller curriculum recommendations. The MSP panel recommendations have incorporated these principles (although the Hilton panel's stress on differential equations has been diminished). The MSP panel strongly endorses the Hilton report's emphasis on the importance within mathematics departments of proper attitudes towards the uses and users of mathematics and of a unified view that respects the content and teaching of pure and applied mathematics equally.

While CUPM and the Hilton panel have been recommending changes in the collegiate mathematics program, the National Council of Teachers of Mathematics has been assessing priorities in school mathematics. The 1980 NCTM booklet, *An Agenda for Action*, recommends "that problem solving be the focus of school mathematics in the 1980s ... that basic skills in mathematics be defined to encompass more than computational facility." Recent nation-wide mathematics tests administered to students in several grades showed uniformly poor performance on questions of a problem solving or application nature. Inevitably these mathematical weaknesses will become more of a problem with college students.

The tentative MSP ideas for curriculum revision were discussed by panel members at sectional and national MAA meetings, at the PRIME 80 Conference, and individually with dozens of mathematics department chairpersons. The helpful criticisms received on these occasions played a vital role in shaping the panel's thinking. It should be noted that several people warned that a mathematical sciences major was unworkable because of the diversity of techniques and modes of reasoning in the mathematical sciences today. Others stated that student course preferences had already "redefined" the mathematics major along the lines being proposed by the MSP panel.

Curriculum Background

Many students today start mathematics in college at a lower level and yet have specific (but uninformed) career goals that require a broad scope of new topics of varying mathematical sophistication. Student changes are reflected in recent upper-level enrollment shifts and the explosion of new theory and applications in all the mathematical sciences. Uncertainties in curriculum produced by these developments have led the MSP panel to look for guidance from past CUPM curriculum development experiences and, farther back, from the traditional goals of the mathematics major before CUPM's creation. No matter how great the advances in the past generation, the traditional intellectual objectives of training in mathematics, defined over scores of years, should be the basis of any mathematical sciences program.

Until the 1950s, mathematics departments were primarily service departments, teaching necessary skills to science and engineering students and teaching mathematics to most students solely for its liberal-arts role as a valuable intellectual training of the mind. The average student majoring in mathematics at a better college in the 1930s took courses in trigonometry, analytic geometry, and college algebra (including calculus preparatory work on series and limits) in the freshman year followed by two years of calculus. While this program may today seem to have unnecessarily delayed calculus, and subsequent courses based on calculus, it did provide students with a background that permitted calculus to be taught in a more rigorous (i.e., more demanding) fashion than it is today.

The mathematics major was filled out with five or six electives in subjects such as differential equations (a second course), projective geometry, theory of equations, vector analysis, mathematics of finance, history of mathematics, probability and statistics, complex analysis, and advanced calculus. Most mathematics majors also took a substantial amount of physics. Training of secondary school mathematics teachers rarely included more than a year of calculus. In the early 1950s, twenty years later, the situation had changed only a little; top schools did now offer modern algebra and abstract analysis.

In 1953, amid reports of widespread dissatisfaction with the undergraduate program, the Mathematical Association of America formed the Committee on Undergraduate Program (CUP, later to be renamed CUPM). CUPM concentrated initially on a unified introductory mathematics sequence Universal Mathematics, consisting of a first semester analysis/college algebra course (finishing with some calculus) followed by a semester of "mathematics of sets" (discrete mathematics). CUPM hoped its Universal Mathematics would "halt the pessimistic retreat to remedial mathematics ...(and)...modernize and upgrade the curriculum."

The first comprehensive curriculum report of CUPM, entitled Pregraduate Training for Research Mathematicians (1963), outlined a model program designed to prepare outstanding undergraduates for Ph.D. studies in mathematics. Emphasis on Ph.D. preparation represented a major departure from the traditional mathematics program and was the source of continuing controversy. A more standard mathematics major curriculum was published in 1965 (revised in 1972), but many colleges also found it to be too ambitious for their students.

For a fuller history of CUPM, see the article of W. Duren (founder of CUPM), "CUPM, The History of an Idea," *Amer. Math. Monthly* 74 (1967), pp. 22-35.

Current Issues

In 1970, 23,000 mathematics majors were graduated. The numbers of Bachelors, Masters, and Doctoral graduates in mathematics had been doubling about every six years since the late 1950s. The 1970 CBMS estimate for the number of Bachelors graduates in mathematics in 1975 was 50,000, but by the late 1970s only 12,000 were graduating annually. Enrollments in many upper-level pure mathematics courses declined even more dramatically in the 1970s as students turned to applied and computer-related courses.

Yet while the number of mathematics majors is decreasing, the demand for broadly-trained mathematics graduates is increasing in government and industry. Mathematical problems inherent in projects to optimize the use of scarce resources and, more generally, to make industry and government operations more efficient guarantee a strong future demand for mathematicians. These problems require people who, fore-

most, are trained in disciplined logical reasoning and, secondarily, are versed in basic techniques and models of the mathematical sciences. In Warren Weaver's words, these are problems of "organized complexity" as well as well-structured applied mathematics of the physical sciences. If mathematics departments do not train these quantitative problem-solvers, then departments in engineering and decision sciences will.

The unprecedented growth of computer science as a major new college subject parallels the theoretical growth of the discipline and its ever-expanding impact on business and day-to-day living. The number of computer science majors now substantially exceeds the number of mathematics majors at most schools offering programs in both subjects. However, computer science has not "taken" students from mathematics, any more than science and engineering take students from mathematics. Rather, computers have generated the need for more quantitative problem-solvers, as noted above.

The shortage of secondary school mathematics teachers nation wide has become worse than ever before. This shortage appears to be due in large measure to the greater attractiveness of computing careers to college mathematics students (indeed high-paying computer jobs are currently luring many teachers out of the classroom). Although the training of future teachers should include course work in computing and applications, such course work heightens the probability that these students will switch to careers in computing.

On another front, pre-calculus enrollments have soared as the mathematical skills of incoming freshmen have been declining (a problem that concerned CUP in its first year). The mathematics curriculum may soon need to allow for majors who do not begin calculus until their sophomore year, as was common a generation ago.

At universities, the decline in graduate enrollments has frequently over-shadowed the decline in undergraduate majors. Faced with heavy precalculus workloads, shrinking graduate programs, and competition from other mathematical sciences departments, university mathematics departments appear less able to broaden and restructure the mathematics major than most liberal-arts college mathematics departments. Many university mathematicians prefer to retain their current pure mathematics major for a small number of talented students.

There are also several encouraging developments. A natural evolution in the mathematics major is occurring at many schools. Students and faculty have developed an informal "contract" for a major that includes traditional core courses in algebra and analysis along with electives weighted in computing and applied mathematics (a formal "contract" major at one school is discussed below).

Another important development is the emphasis on systems design, as opposed to mathematical computation, in current computer science curricula. The Association for Computing Machinery Curriculum 78 Report delegates the responsibility for teaching numerical analysis, discrete structures, and computational modeling to mathematics departments. This ACM curriculum implicitly encourages students interested in computer-based mathematical problem solving to be mathematical sciences majors. The MSP panel has been careful to coordinate its work with computer science curriculum groups in order to minimize potential conflicts and maximize compatibility between computer science and mathematical sciences programs.

Curricular Principles

The goal of this panel was to produce a flexible set of recommendations for a mathematical sciences major, a major with a broad, historically rooted foundation for dealing with current and future changes in the mathematical sciences. The panel sought a unifying philosophy for diverse course work in analysis, algebra, computer science, applied mathematics, statistics, and operations research.

Program Philosophy

I. The curriculum should have a primary goal of developing attitudes of mind and analytical skills required for efficient use and understanding of mathematics. The development of rigorous mathematical reasoning and abstraction from the particular to the general are two themes that should unify the curriculum.

II. The mathematical sciences curriculum should be designed around the abilities and academic needs of the average mathematical sciences student (with supplementary work to attract and challenge talented students).

III. A mathematical sciences program should use interactive classroom teaching to involve students actively in the development of new material. Whenever possible, the teacher should guide students to discover new mathematics for themselves rather than present students with concisely sculptured theories.

IV. Applications should be used to illustrate and motivate material in abstract and applied courses. The development of most topics should involve an interplay of applications, mathematical problem-solving, and theory. Theory should be seen as useful and enlightening for all mathematical sciences.

V. First courses in a subject should be designed to appeal to as broad an audience as is academically reasonable. Many mathematics majors do not enter college planning to be mathematics majors, but rather are attracted by beginning mathematics courses. Broad introductory courses are important for a mathematical sciences minor.

Course Work

VI. The first two years of the curriculum should be broadened to cover more than the traditional four semesters of calculus-linear algebra-differential equations. Calculus courses should include more numerical methods and non-physical-sciences applications. Also, other mathematical sciences courses, such as computer science and applied probability and statistics, should be an integral part of the first two years of study.

VII. All mathematical sciences students should take a sequence of two upper-division courses leading to the study of some subject(s) in depth. Rigorous, proof-like arguments are used throughout the mathematical sciences, and so all students should have some proof-oriented course work. Real analysis or algebra are natural choices but need not be the only possibilities. Proofs and abstraction can equally well be developed through other courses such as applied algebra, differential equations, probability, or combinatorics.

VIII. Every mathematical sciences student should have some course work in the less theoretically structured, more combinatorially oriented mathematics associated with computer and decision sciences.

IX. Students should have an opportunity to undertake "real-world" mathematical modeling projects, either as term projects in an operations research or modeling course, as independent study, or as an internship in industry.

X. Students should have a minor in a discipline using mathematics, such as physics, computer science, or economics. In addition, there should be sensible breadth in physical and social sciences. For example, a student interested in statistics might minor in psychology but also take beginning courses in, say, economics or engineering (heavy users of statistics).

Building Mathematical Maturity

As noted in Principle I, a major in mathematical sciences should emphasize general mathematical reasoning as much as mastery of various subject matter. Implicit in this principle is that less material would be covered in many courses but that students would be expected to demonstrate a better understanding of what is taught, e.g., by solving problems that require careful mathematical analysis.

This mathematical sciences curriculum would model the historical evolution of mathematical subjects: some problems are introduced, formulas and techniques are developed for solving problems (usually with heuristic explanations), then common aspects of the problems are examined and abstracted with the purpose of better understanding "what is really going on." The difference in this scheme between beginning calculus and upper-division probability theory would be primarily a matter of the difficulty of the problems and techniques and the speed with which the material is covered and generalized, i.e., a matter of the mathematical maturity of the audience. In the course of two or three years of such course work, there would be a steady increase in sophistication of the material and more importantly, an increase in the student's ability to learn and organize the ideas of a new mathematical subject. Students should be able to read and learn mathematics on their own from texts. The MSP panel feels that such maturity is a function of how a subject is learned as much as what is learned.

All courses should have some proofs in class and, as the maturity of students increases, occasional proofs as homework exercises. In particular, students should acquire facility with induction arguments, a basic method of proof in the mathematical sciences. After reviewing performances of current students and programs of mathematics students 30 years ago, the MSP panel has concluded that many able students do not now have, nor were they previously expected to have, the mathematical maturity to take theoretical courses before their senior year. On the other hand, by the senior year, all students should be ready for some proof-oriented courses that show the power of mathematical abstraction in analyzing concepts that underlie a variety of concrete problems. For example, part of a flowchart of courses leading to a senior-year real analysis course

might be:

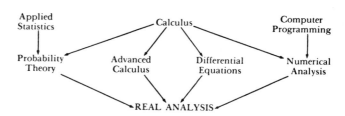

Core Requirements

The panel has found the question of whether to require courses in algebra and analysis its most controversial problem. In light of the strongly differing opinions received on this subject, the MSP panel is making only a minimal recommendation (Principle VII) that it feels is reasonable for all students. Possible two course sequences besides a year of analysis or of algebra are: analysis and proof-oriented probability theory, analysis and differential equations, abstract algebra and (proof-oriented) combinatorics, applied algebra and theory of computation, or analysis and a topics-in-analysis seminar. While not a sequence, one course in analysis and one course in algebra also fulfill the spirit of this requirement. Some departments will want to make stronger requirements. The issue of theory requirements is discussed more fully below.

Students should not be required to study a subject with an approach whose rationale depends on material in later courses nor should they be required to memorize (blindly) proofs or formulas. Some upper-level elective courses should always be taught as mathematics-for-its-own-sake, but an instructor should be very careful not to skip the historical motivation and application of a subject in order to delve further into its modern theory.

The recommendation for interactive teaching (Principle III) seeks to encourage student participation in developing new mathematical ideas. It constrains an instructor to teach at a level that students can reasonably follow. Interactive teaching implicitly says that mathematics is learned by actively doing mathematics, not by passively studying lecture notes and mimicking methods in a book. Without needlessly slowing progress in class, an instructor should discuss how one can learn much from wrong approaches suggested by students. New mathematical theories are not divined with textbook-like compact proofs but rather involve a long train of trial-and-error creativity.

Henry Pollak expressed this need in the Conference Board of Mathematical Sciences book, *The Role of Ax-*

iomatics and Problem Solving in Mathematics (Ginn, 1966):

> A carefully organized course in mathematics is sometimes too much like a hiking trip in the mountains that never leaves the well-constructed trails. The tour manages to visit a steady sequence of the high spots in the natural scenery. It carefully avoids all false starts, dead ends and impossible barriers and arrives by five o'clock every afternoon at a well-stocked cabin. ...However, you miss the excitement of occasionally camping out or helping to find a trail and of making your way cross-country with only a good intuition and a compass as a guide. "Cross country" mathematics is a necessary ingredient of a good education.

Further details about the course work recommendations in Principles VI, VIII, and IX appear in later chapters of this report. Discussion of courses in discrete methods, applied algebra, and numerical analysis appears in the last section of this chapter.

Teaching Mathematical Reasoning

Because a mathematical sciences major must include a broader range of courses than a standard (pure) mathematics major, many mathematicians have expressed concern that it will be harder to teach the average mathematics student rigorous mathematical reasoning in a mathematical sciences major. They believe that the major will develop problem-solving skills but that without more abstract pure mathematics, students will never develop a true sense of rigorous mathematical reasoning. The MSP panel thinks that a mathematical sciences major with a strong emphasis on problem-solving is in keeping with time-tested ways of developing "mathematical reasoning." The question of whether to require "core" pure mathematics courses, such as abstract algebra and real analysis, in any mathematical sciences major is discussed in the next section.

Historically (before 1940), the main thrust of the mathematics major at most colleges was problem-solving. Most courses in the major could be classed as mathematics for the physical sciences: trigonometry, analytic geometry, calculus (first-year and advanced), differential equations, and vector analysis. Proofs in advanced calculus were symbolic computations. Proofs in number theory were, and still are, usually combinatorial problems. The one abstract "pure" course in the curriculum was logic. A "rigorous" course did not mean an abstract course, "mathematics done right." A rigorous course used to mean a demanding, more in-depth treatment that required more skill and ingenuity from the student. The past curriculum surely had some faults, but its problem-solving and close ties to physics

came from traditions that go back to the roots of mathematics.

While problem solving may traditionally be the primary way of teaching mathematical reasoning to undergraduates, the complexity and breadth of modern mathematics and mathematical sciences require theory to help organize and simplify learning. Rigorous problem solving should lead students to appreciate theory and formal proofs. In a mathematical sciences major, theory should be primarily theory for a purpose, theory born from necessity (of course, this is also the historical motivation of most theory). Students may find theory difficult, but they should never find it irrelevant.

Most courses in a mathematical sciences major should be case studies in the pedagogical paradigm of real world questions leading to mathematical problem solving of increasing difficulty that forces some abstraction and theory. As mentioned earlier, lower-level courses would concentrate on problem solving to build technical skills with occasional statements of needed theorems, while typical upper-level courses would concentrate on problem solving to build technical skills with occasional statements of needed theorems, while typical upper-level courses would emphasize the transition from harder problem-solving to theory.

Instructors should resist pressures to survey fully fields such as numerical analysis, probability, statistics, combinatorics, or operations research in the one course a department may offer in the field. The instructor of such a course should give students a sense of the problems and modes of reasoning in the field, but after that, should be guided by the pedagogical model given above. All syllabi produced by MSP subpanels should be viewed in this light. Most instructors will cover most of a suggested syllabus, but general pedagogical goals should always take precedence over the demands of individual course syllabi.

The MSP panel believes that for generations mathematics instructors have used the paradigm mentioned above to develop rigorous mathematical reasoning. Implicit in this paradigm is a unity of purpose between students and instructor. Most students like to start with concrete real-world examples as a basis for mathematical problem solving. They expect the problems to get harder and require more skill and insight. And they certainly appreciate theory when it makes their work easier (although understanding formal proofs of useful theory requires maturity). Interactive teaching also becomes natural: students are interested in participating in a class that is developing a subject in a way that they can appreciate.

How Much Theory?

This section summarizes arguments for and against requiring upper-level analysis and algebra courses of all mathematical sciences majors, and why the MSP panel made its "compromise" decision.

Expecting controversy on several issues, the MSP panel organized sessions at national and regional MAA meetings to get input from the mathematics community. The main area of contention was how many courses to require in specific areas. The panel heard complaints that some areas were being neglected or that only one course in a certain area would be so superficial as to be worse than no course. However, most constituencies came to accept the need for compromise recommendations of limited exposure to several areas with students left to choose for themselves an area to study in greater depth. On the other hand, one important issue emerged on which a compromise position seemed to antagonize at least as many people as it pleased. This was the question of whether to require an analysis and/or an abstract algebra course and, more generally, how much proof-oriented course work should be required in a mathematical sciences major.

In the early 1970's, a majority of mathematics programs required at least these two upper-level "core mathematics" courses for all students. Recently, declining enrollments in these courses and student preference for more applied or computing courses have forced many departments either to relax this requirement or to introduce a new applied track which does not require these two courses. People favoring the requirement of analysis and algebra argue that:

- Not requiring them would speed an already dangerous deterioration in the intellectual basis of the mathematics major;
- A major without at least analysis and algebra would be a superficial potpourri of courses—a major of no real value to anyone, e.g., graduate study in statistics requires analysis and (linear) algebra;
- One cannot understand "what mathematics is about" without these two courses—a major without these two courses simply should not be offered by a mathematics department.

People in favor of not requiring analysis and algebra argue that:

- With a more applied emphasis the mathematical sciences major will attract more good students, whereas requiring these courses would mean no change (except for new applied electives) from the 1960s type of mathematics major that today attracts only a marginal number of students;

- Analysis and algebra are fine for some students but demand a mathematical maturity that many other undergraduates lack—these students memorize proofs blindly to pass examinations and never take the follow-on courses needed to appreciate the structure and elegance of these subjects; and
- Proofs and abstraction can equally well be developed through other courses such as applied algebra, probability, differential equations, or combinatorics.

Mathematicians must face the reality of a general change in the attitude of college students towards mathematics. The popularity of science and mathematics in the 1960s drew more of the brightest students to mathematics and also motivated all students to work harder at mathematics in high school. So the average mathematics student was capable of handling a more theoretical mathematics program.

Today, mathematics appears to be getting no more than its traditional (smaller) share of bright students, and high school study habits are less good. However, almost all of today's mathematics students still find a few subjects, pure or applied, particularly interesting and want to study this material in some depth. Also by the senior year, the MSP panel believes that mathematics majors do have the mathematical maturity to appreciate, say, a moderately abstract real analysis course. Examples of new approaches to teaching analysis and other core mathematics courses appear in subsequent chapters.

Since there was agreement on the importance of some theoretical depth, the MSP panel proposed the compromise of Principle VII, recommending "a sequence of two upper-division courses leading to the study of some subject in depth." Because of the lack of consensus on the analysis-algebra question, the MSP panel expects this issue to be debated and modified at individual institutions. The faculty should not require courses that most students strongly dislike, nor should faculty shy away from any theory requirements for fear of losing majors. The faculty rather must motivate students to appreciate the value of some theoretical course work.

Sample Majors

This section presents two 12 semester-course mathematical sciences majors. Many other sample majors could be given. The MSP panel believes that most majors should be a "convex combination" of the two majors given here. Major A contains much of a standard mathematics major, while Major B is a broader program designed for students interested in problem solving. Both majors should be accompanied by a minor in a related subject.

The common core of all majors would be three semesters of calculus, one course in linear algebra, one course in computer science plus either a second computer course or extensive use of computing in several other courses, one course in probability and statistics, the equivalent of a course in discrete methods, modeling experience, and two theoretical courses of continuing depth.

Mathematical Sciences Major A

- Three semesters of calculus
- Linear algebra
- Probability and statistics
- Discrete methods
- Differential equations (with computing)
- Abstract algebra (one-half linear algebra)
- Two semesters of advanced calculus/real analysis
- One course from the following set: abstract algebra (second course), applied algebra, geometry, topology, complex analysis, mathematical methods in physics
- Mathematical modeling
- Plus related course work: two semesters of computer science and two semesters of physics, to be taken in the first two years.

Mathematical Sciences Major B

- Three semesters of calculus
- Linear algebra
- Introduction to computer science
- Numerical analysis *or* second course in computer science
- Probability and statistics
- Advanced calculus *or* abstract algebra
- Discrete methods *or* differential equations
- Mathematical modeling or operations research
- Two electives continuing a subject with theoretical depth.

Subsequent sections in this report contain recommendations for discrete methods, applied algebra, and numerical analysis courses; for calculus, linear algebra, and differential equations courses; for upper-level core mathematics; for computer science; for modeling and operations research; and for probability and statistics.

Major A is meant to be close to the spirit of the major suggested by the NRC Panel on Training in Applied Mathematics. That panel viewed differential equations as a unifying theme in the major. The proper mixture of Majors A and B (with appropriate electives) would also allow students to make statistics or operations research a unifying theme.

The MSP panel feels that a set of courses similar to either of the above two majors, or a mixture thereof, would be reasonable for most mathematical sciences students. Some departments could offer several tracks for the mathematical sciences major. Special areas of faculty strength or student interest should obviously be reflected in the curriculum.

Computing assignments should be used in most courses. When a liberal arts college mathematics department teaches computer science, such computing course work must frequently be counted within the college limit of 12 or 13 courses permitted in one department. This regulation is assumed in Major B. However, the MSP panel believes that counting computer courses this way unfairly restricts a mathematical sciences major. One alternative is to list computer courses through the Computing Center.

The one fundamental new course in these sample majors is discrete methods. As mentioned in Principle VIII, the MSP panel feels that the central role of combinatorial reasoning in computer and decision sciences requires that some combinatorial problem solving should be taught in light of the three semesters devoted to analysis-related problem solving in the calculus sequence. To this end, the modeling course should be heavily combinatorial if students have not taken a formal discrete methods course.

Major A would be good preparation for graduate study in mathematics, applied mathematics, statistics, or operations research as well as many industrial positions as a mathematical analyst or programmer. Major B would be good preparation for most industrial positions and for graduate study in applied mathematics, statistics, or operations research (for such graduate study, both advanced calculus and upper-level linear algebra are usually needed). Representatives from many good mathematics graduate programs have stated that they would accept strong students with Major B-type training.

Many computer science graduate programs would accept Major B if the two electives were in computer science (although some other undergraduate computer science course deficiencies may still have to be made up in the first year of graduate study). In a computer science concentration within a mathematical sciences major, modern algebra might be replaced by applied algebra (see below for more details). Major B with an elective in the theory of interest and a second probability-statistics course would be excellent preparation for actuarial careers. Students interested in physical sciences-related applied mathematics could modify either sample major to get a good program. Both majors provide preparation for secondary school mathematics teaching, when supplemented with teaching methodology and practicum courses (theory courses must include algebra and geometry).

Many smaller schools are being forced to offer a program in the spirit of Major B because almost all of B's courses have the needed enrollment base of students drawn from outside mathematics.

The courses involving numerical analysis, probability and statistics, discrete methods, and modeling all can be designed as lower-level or upper-level courses. A large amount of flexibility is possible in "repackaging" the mathematical sciences material. For example, a Computational Models course (see the 1971 CUPM *Report on Computational Mathematics*) could cover some numerical analysis along with a little applied probability and statistics to be used in simulation modeling. A quarter system institution would have even greater flexibility in implementing this major.

Mathematical Sciences Minor

Just as a mathematical sciences major should be accompanied by a minor in a related subject, so also do many other disciplines encourage their students to have a minor, or double major, in mathematics. At some colleges, as many as half the mathematics majors have another major. Unfortunately, while mathematical methods are playing an increasingly critical role in social and biological sciences and in business administration, students are generally ignorant or misinformed in high school and early college years about the importance of mathematics in these areas.

The result is that many students either do not realize the value of further course work in the mathematical sciences until their junior or senior year, or their poor high school preparation forces them to take a year of remedial mathematics before they can begin to learn any of the college mathematics they need. For such students, a traditional six to eight course minor in mathematics, starting with (at least) three semesters of calculus, is not feasible. When students in the social and biological sciences come to realize the value of mathematics in the junior year, they have frequently had only one semester of calculus, or perhaps a year of calculus with probability.

The MSP panel believes that these students would be well served by a six to eight course mathematical sciences minor consisting of two semesters of calculus, one semester of (calculus-based) probability and statistics,

one semester of introductory computer science, plus two to four electives chosen from courses such as numerical analysis, discrete methods, linear algebra, differential equations, linear programming, mathematical modeling, and additional courses in calculus, probability or statistics, and computer science. Such a minor could easily be completed in three semesters. It has little prerequisite structure so that students can immediately pick courses based on personal interests rather than initially "mark time" waiting to complete the calculus sequence.

Such a minor has several important points in its favor. First of all, this minor is a collection of useful mathematical sciences courses which present concepts and techniques that arise frequently in the social and biological sciences. While this minor lacks the mathematical depth of the traditional type of mathematics minor, it nonetheless introduces students to important modes of mathematical reasoning. Second, such a minor will be attractive to students because it enhances employment opportunities and prospects for admission to graduate or professional schools. Third, after the exposure to interesting mathematical sciences topics, some students will want to study these subjects further in graduate school, either in a mathematical sciences graduate program or as electives in other graduate programs. Fourth, this minor will bring more students into mathematical sciences courses, making it possible to offer these courses more frequently. Conversely, offering more mathematical sciences courses each semester will make a mathematical sciences minor, as well as the regular mathematical sciences major, more attractive to students. In addition, when more students are taking mathematical sciences courses and finding out how useful mathematics is, the campus-wide student awareness of the value of mathematics will increase.

Examples of Successful Programs

Proper curriculum is the heart of a mathematical sciences program, but there are many non-academic aspects that also must be considered. A wide variety of course offerings is not as important as the spirit with which the general program is offered. This section discusses salient features of some successful mathematics programs. "Successful" means attracting a large number of students into a program that develops rigorous mathematical thinking and also offers a spectrum of (well taught) courses in pure and applied mathematics. Successful programs typically produce 5% to 8% of their college's graduates, although nation wide, mathematics majors constitute only about 1% of college graduates. Faculty and student morale is uniformly high in these programs. As one would expect, teaching and related student-oriented activities consume most of the faculty's time in such successful programs, and there is little faculty research. The professors' pride in good teaching and in the successes of their students leaves them with few regrets about not publishing. The set of programs mentioned here is only a sampling of successful programs that have come to the attention of this CUPM panel. More detailed information about these mathematics programs is available from individual colleges.

Saint Olaf College, a 2800-student liberal arts college in Northfield, Minnesota, has a contract mathematics major. Each mathematics student presents a proposed contract to the Mathematics Department. The contract consists of at least nine courses (college regulations limit the maximum number of courses that can be taken in one department to 14). The department normally will not accept a contract without at least one upper-level applied and one upper-level pure mathematics course, a computing course or evidence of computing skills, and some sort of independent study (research program, problem-solving proseminar, colloquium participation, or work-study).

Frequently a student and an advisor will negotiate a proposed contract. For example, a faculty member will try to persuade a student interested in scientific computing and statistics that some real analysis and upper-level linear algebra should be included in the contract by showing that this material is needed for graduate study in applied areas, and in any case a liberal arts education entails a more broadly based mathematics major. Conversely, a student proposing a pure mathematics contract would be confronted with arguments about not being able to appreciate theory without knowledge of its uses. In the end, the student and the faculty member understand and respect each other's point of view.

This understanding of each other's interests naturally carries into the classroom. Also, the contract negotiations "break the ice" and make students more at ease in talking to faculty (and encourage constructive criticism). The Mathematics Department offers minors in computing and statistics, but the attractiveness of a contract major in mathematics leads most students interested in these areas eventually to become mathematics majors.

Lebanon Valley College, a small (1000-student) liberal arts college in Pennsylvania, has only five mathematics faculty but its Department of Mathematical Sciences offers majors in Mathematics, Actuarial Science, Computer Science, and Operations Research. The course work in the mathematics graduate preparation track involves a problem seminar, Putnam team sessions, and formal and informal topics courses (because of the limited demand in this area). All mathematical sciences majors must take a rigorous 25 semester-hour core of calculus, differential equations, linear algebra, foundations, and computer science. Most courses are peppered with applications and computing assignments.

The mathematics faculty are heavily involved in recruiting students by attending College Fairs and College Nights and by visiting regional high schools to explain to students and counselors the many diverse and attractive careers in the mathematical sciences, and the importance of mathematics in other professions. As a result of this effort, 10% of the incoming Lebanon Valley freshmen plan majors in the mathematical sciences (the national average is 1%), and 7% of Lebanon Valley graduates are mathematical sciences majors. Many students are initially attracted by the major in actuarial science (an historically established profession) and then move into other areas of applied and pure mathematics, but this pattern may change with the newly established computer science major.

Once the faculty have the "students' attention," they work the students hard. The students respond positively to the demands of the faculty for three reasons. First, known rewards await those who do well in mathematics (besides the obvious long-term rewards, the department awards outstanding students with membership in various professional societies in the mathematical sciences). Second, a personal sense of intellectual achievement is carefully nurtured starting in the freshman year with honors calculus for mathematics majors. Finally, as at St. Olaf, a continuing dialogue between students and faculty allows students to help shape the mathematics program. In fact, students interview candidates for faculty positions and their recommendations carry great weight. The department keeps in close touch with alumni by sending each one a personal letter every other year with news about the department and fellow alumni.

Nearby Gettysburg College has a special vitality in its mathematics program that comes from an interdisciplinary emphasis. The department has held joint departmental faculty meetings with each natural and social science department at Gettysburg to discuss common curriculum and research interests. Several interdisciplinary team-taught courses have been developed, such as a course on symmetry taught jointly by a mathematician and a chemist. An interdepartmental group organized two recent summer workshops in statistics which drew faculty from eight departments. Mathematics faculty have audited a variety of basic and advanced courses in related sciences to learn to talk the language of mathematics users. Mathematics faculty bring this interdisciplinary point of view into every course they teach, giving interesting applications and showing, say, how a physicist would approach a certain problem. Needless to say, a large number of mathematics majors at Gettysburg are double majors.

Frequently a separate computer science department with its own major spells disaster for the mathematics major at a college. But Potsdam State College (in the economically depressed northeast corner of New York) has possibly the greatest percentage of mathematics graduates of any public institution in the country—close to 10%—despite competition from a popular computer science major. The most striking feature to a visitor to the Potsdam State Mathematics Department is the great enthusiasm among the students and the sense of pride students have in their ability to think mathematically. (While it is hard to measure objectively these students' mathematical development, leading technological companies, such as Bell Labs, IBM, and General Dynamics, annually hire several dozen Potsdam mathematics graduates.)

Classes have a limited amount of formal lectures. Most time is spent discussing work of the students. The emphasis on giving students a sense of achievement is due in large part to experiences of the Potsdam chairman when he taught in a Black southern institution. By instilling self confidence, he had helped able but ill-prepared students excel in calculus and even saw some go on to good mathematics graduate programs. The department has various awards for top students, a very active Pi Mu Epsilon chapter, publications about careers in mathematics and successes of former students, and a large student-alumni newsletter. Upper-class mathematics students are used to tutor (and encourage) beginning students. They also communicate their enthusiasm about mathematics to friends and teachers back home. As a result, half the incoming Potsdam freshmen sign up for calculus (although few departments require it).

The computer science major at Potsdam State is viewed by the mathematics faculty as a great asset to the Mathematics Department. The computer science major helps attract good students to Potsdam who often decide to switch to, or double major with, mathematics. Also the computer science program offers career skills and needed mathematical breadth. Numerical analysis, operations research, and modeling are taught in computer science (the Mathematics Department has

had to limit severely their upper-level electives in order to keep class size down and preserve small group seminars).

As noted at the start of this section, the preceding mathematics programs represent only a small sampling of the excellent programs in this country. Several women's colleges offer fine programs worth noting. For example, the Goucher College Mathematics Department has integrated computing in almost all courses and has a broad curriculum in pure and applied mathematics; and the Mills College Mathematics Department has successfully promoted the critical role of mathematics for careers in science and engineering. The cornerstone of Ohio Wesleyan's excellent mathematics program is an innovative calculus sequence (with computing, probability, and diverse mathematical modeling). Georgia State University, an urban public institution with a highly vocational orientation, has a Mathematics Department that has broken out of the typical low-level service function mode to offer a fine, well-populated mathematical sciences major. While research and graduate programs often dominate concerns about the undergraduate mathematics major at universities, mathematics faculty at many universities work closely with undergraduate majors in excellent unified mathematical sciences programs. Three such institutions are Clemson University, Lamar University (Texas), and Rensselaer Polytechnic Institute.

Most universities today have separate departments in computing and mathematical sciences. To counter this division, the University of Iowa and Oregon State University have developed unified inter-departmental mathematical sciences majors. The MSP panel strongly endorses such inter-departmental majors. At some universities, most of the mathematical sciences, outside of pure mathematics, have been housed in one department. Although the MSP panel prefers a unified mathematical sciences major (ideally in one department), several of these non-pure mathematical sciences departments have good undergraduate programs that may be of interest to other institutions: the Mathematical Sciences Department at Johns Hopkins University, the Mathematical Sciences Department at Rice University, and the Department of Applied Mathematics and Statistics at the State University of New York at Stony Brook.

Departmental Self-Study and Publicity

The MSP panel urges all mathematics departments to engage in serious self-study to identify one or more major themes to emphasize in their mathematical sciences programs: an interdisciplinary focus in cooperation with other departments; an innovative calculus sequence (integrating computing, applications, etc.); a work-study program or other individualized learning experience; special strength in one area of the mathematical sciences (pure or applied); or a track directed towards employment in a regional industry (such as aerospace, automative, insurance). Some colleges have successfully developed a multi-option major, but usually such programs are the outgrowth of successful one-theme programs that slowly added new options (for example, the multiple-major mathematical sciences program at Lebanon Valley College, mentioned in the preceding section, started with just an Actuarial Science option). The MSP panel's advice is first to do one thing well.

A departmental emphasis should be consistent with the general educational purposes of the whole institution and the academic interests of the high school graduates who have historically gone to that institution. It is very risky to design a mathematical sciences program about a theme that the mathematics faculty find attractive and then to try to recruit a new group of high school students to come to the institution for this program. Note that a thematic emphasis does not mean that basic parts of the mathematical sciences program discussed earlier in this chapter can be neglected.

Following a departmental self-study and implementation of its recommendations for new courses or development of industrial work-study contacts, etc., it is next necessary to publicize the mathematics department's program with brochures and visits to regional high schools and College Fairs. Virtually all mathematics departments with large programs (where mathematical sciences majors constitute over 4% of the school's graduates) have extensive publicity programs. Such publicity should emphasize the general usefulness of mathematics in the modern world, whether a student is a prospective mathematical sciences major or minor or an undecided liberal arts student.

High school guidance counselors often do not realize that there are other attractive mathematics-related careers outside straight computing. Counselors tend to be afraid of mathematics because of their own personal difficulties with the subject. Some counselors have been known to discourage students from taking more than the minimum required amount of high school mathematics with the warning that students risk getting poor grades in (hard) mathematics courses and thus hurting their chances of college admission.

College faculty trying to publicize the value of mathematics and its study at their institution should seek the cooperation of local associations of the National Council of Teachers of Mathematics, which have long been working to promote interest in mathematics in the high schools.

New Course Descriptions

Finite structures are used throughout the mathematical sciences today. Two new basic courses about finite structures belong in the mathematical sciences curriculum, one addressing combinatorial aspects and one addressing algebraic aspects. Another topic, numerical analysis, has become more important with the growth of computer science. This section describes a numerical analysis course that is more applied and at a lower level than the previous CUPM numerical analysis recommendations (Course 8 in the CUPM report *A General Curriculum for Mathematics in Colleges.*)

Discrete Methods Course

This course introduces the basic techniques and modes of reasoning of combinatorial problem solving in the same spirit that calculus introduces continuous problem solving. The growing importance of computer science and mathematical sciences such as operations research that depend heavily on combinatorial methods justifies at least one semester of combinatorial problem solving to balance calculus' three semesters of analysis problem solving.

Unlike calculus, combinatorics is not largely reducible to a limited set of formulas and operations. Combinatorial problems are solved primarily through a careful logical analysis of possibilities. Simple ad hoc models, often unique to each different problem, are needed to count or analyze the possible outcomes. This need to constantly invent original solutions, different from class examples, is what makes the discrete methods course so valuable for students.

Like calculus, combinatorics is a subject which has a wide variety of applications. Many of them are related to computers and to operations research, but others relate to such diverse fields as genetics, organic chemistry, electrical engineering, political science, transportation, and health science. The basic discrete methods course should contain a variety of applications and use them both to motivate topics and to illustrate techniques.

The course has an enumeration part and a graph theory part. These parts can be covered in either order. While texts traditionally do enumeration first, the graph material is more intuitive and hence it seems natural to do graph theory first (as suggested below).

With the right point-of-view, many combinatorial problems have quite simple solutions. However, the object of this course is not to show students simple answers. It is to teach students how to discover such simple answers (as well as not so simple answers). The means for achieving solutions are of more concern than the ends. Learning how to solve problems requires an interactive teaching style. It requires extensive discussion of the logical faults in wrong analyses as much as presenting correct analyses.

Since the course should emphasize general combinatorial reasoning rather than techniques, a large degree of flexibility is possible in the choice of topics. The course outline given below contains many optional topics. Some of the core topics, such as the inclusion-exclusion formula, might also be skipped to allow the course to be tailored to the interests of students.

COURSE OUTLINE

I. Graph Theory
   A. *Graphs as models.* Stress many applications.
   B. *Basic properties of graphs and digraphs.* Chains, paths, and connectedness; isomorphism; planarity.
   C. *Trees.* Basic properties; applications in searching; breadth-first and depth-first search; spanning trees and simple algorithms using spanning trees. Optional: branch and bound methods; tree-based analysis of sorting procedures.
   D. *Graph coloring.* Chromatic number; coloring applications; map coloring. Optional: related graphical parameters such as independent numbers.
   E. *Eulerian and Hamiltonian circuits.* Euler circuit theorem and extensions; existence and non-existence of Hamiltonian circuits; applications to scheduling, coding, and genetics.
   F. *Optional topics:*
      a. Tournaments
      b. Network flows and matching
      c. Intersection graphs
      d. Connectivity
      e. Coverings
      f. Graph-based games
II. Combinatorics
   A. *Motivating problems and applications.*
   B. *Elementary counting principles.* Tree diagrams; sum and product role; solving problems that must be decomposed into several subcases. Optional: applications to complexity of computation, coding, genetic codes.

C. *Permutations and combinations.* Definitions and simple counting; sets and subsets; binomial coefficients; Pascal's triangle; multinomial coefficients; elementary probability notions and applications of counting. Optional: algorithms for enumerating arrangements and combinations; binomial identities; combinations with repetition and distributions; constrained repetition; equivalence of distribution problems, graph applications.
D. *Inclusion/exclusion principle.* Modeling with inclusion/exclusion; derangements; graph coloring. Optional: rook polynomials.
E. *Recurrence relations.* Recurrence relation models; solution of homogeneous linear recurrence relations; Fibonacci numbers and their applications.
F. *Optional topics:*
   a. Generating functions
   b. Polya's enumeration formula
   c. Experimental design
   d. Coding

The preceding course outline is for either a one-semester or a two-quarter course. A two-quarter course has a natural structure, covering enumerative material in one quarter and graph theory plus designs in another quarter. There are several books available for part or all of the discrete methods course. It is anticipated that as this discrete methods course becomes more widely taught, many more books will become available and the exact nature of the syllabus will evolve.

There are several obvious places where a computer can be used in this course: ways of representing graphs in a computer and performing simple tests (e.g., connectivity); asymptotic calculations in enumeration problems; network flow algorithm; and algorithms for enumerating permutations and combinations. The pedagogical problem is that computer programming takes time away from problem-solving exercises, possibly too much time if a school's computer operation runs in a batch processing mode.

A more advanced second course in combinatorics may also be considered. This course can treat core topics in the discrete methods course in greater depth, and some of the optional topics. Other important topics are Ramsey theory, matroids, and graph algorithms. The course could concentrate on combinatorics or on graph theory, or could be a topics course which varies from year to year. Some of the texts listed below would be suitable for this second combinatorics course.

Combinatorics & Graph Theory Texts

1. Bogart, Kenneth, *Introductory Combinatorics*, Pitman, Boston, 1983.
2. Brualdi, Richard, *Introductory Combinatorics*, Elsevier-North Holland, New York, 1977.
3. Cohen, Daniel, *Basic Techniques of Combinatorial Theory*, J. Wiley & Sons, New York, 1978.
4. Liu, C.L., *Introduction to Combinatorial Mathematics*, McGraw Hill, New York, 1968.
5. Roberts, Fred, *Applied Combinatorics*, Prentice-Hall, Englewood Cliffs, New Jers., 1984.
6. Tucker, Alan, *Applied Combinatorics*, J. Wiley & Sons, New York, 1980.

Graph Theory Texts

1. Bondy, J. and Murty, V.S.R., *Graph Theory with Applications*, American Elsevier, New York, 1976.
2. Chartrand, Gary, *Graphs as Mathematical Models*, Prindle, Weber, and Schmidt, Boston, 1977.
3. Ore, Oystein, *Graphs and Their Uses*, Math. Assoc. of America, Washington, D.C., 1963.
4. Roberts, Fred, *Discrete Mathematical Models*, Prentice-Hall, Englewood Cliffs, New Jersey, 1976.
5. Trudeau, Robert, *Dots and Lines*, Kent State Press, Kent, Ohio, 1976.

Combinatorics Texts

1. Berman, Gerald and Fryer, Kenneth, *Introduction to Combinatorics*, Academic Press, New York, 1969.
2. Eisen, Martin, *Elementary Combinatorial Analysis*, Gordon-Breach, New York, 1969.
3. Vilenkin, N., *Combinatorics*, Academic Press, New York, 1971.
4. Street, A. and Wallis, W., *Combinatorial Theory: An Introduction*, Charles Babbage, 1975.

Applied Algebra Course

(*Editorial Note in 1989 reprinting*: This course is now called **Discrete Structures** and is usually now taught at the freshman level. The course discussed here is more advanced and intended for the sophomore-junior level.)

A traditional time for an applied algebra course is in the junior year—when students would be ready for a modern algebra course. However, as noted above, many students will not be ready for algebraic abstraction until senior year. The course builds on experiences in beginning computer science courses that have implicitly imparted to students a sense of the underlying algebra of computer science structures, and formally presents topics like Boolean algebra, partial orders, finite-state machines, and formal languages that will be used in later computer science courses. At the same time, this course can also be very rewarding to regular mathematics majors who should appreciate the new algebraic structures such as formal languages and finite state machines that are so different from the structures in the regular abstract algebra course. Substantial class time should be spent on proofs with special emphasis on induction arguments. This course is just as mathematically sophisticated and capable of developing abstract reasoning as abstract algebra, but the topics stress set-relation systems rather than binary-operation systems. Indeed the abstract complexity of the basic structures is much greater in applied algebra, but this complexity precludes the construction of logical pyramids built of simple algebraic inferences common to many areas of abstract algebra.

This course is an advanced version of the lower-division B3 Discrete Structures course in ACM Curriculum 68. The B3 course was the source of much dissatisfaction because it contained a huge amount of material, and it required too great mathematical maturity for a lower-division course. The recent ACM Curriculum 78 recommends that the B3 course be treated as a more advanced course and that it should be taught in mathematics departments rather than computer science departments. The B3 course was the subject of several papers at meetings of the ACM Special Interest Group in Computer Science Education (SIGCSE); see the February issues (Proceedings of SIGCSE annual meeting) of the *SIGCSE Bulletin* in 1973, 1974, 1975, 1976.

The B3 course contained both applied algebra and discrete methods. The MSP panel recommends that a separate full course be devoted to discrete methods (see the discrete methods course description earlier in this Section). Because some computer science courses may devote a substantial amount of time introducing some of the topics in the above applied algebra syllabus, the exact content of this course will vary substantially from college to college. For this reason the syllabus outline was kept brief. At some colleges, applied algebra will still have to be combined with discrete methods in one course (the computer science major may not have the time for two separate courses). The applied algebra part of such a combined course would, in most cases, concentrate on topics 1, 2, 3, 4, 6 in the syllabus. Many of the discrete structures texts listed below cover both applied algebra and discrete methods.

Course Topics

A. Sets, binary relations, set functions, induction, basic graph terminology.

B. Partially ordered sets, order-preserving maps, weak orders.

C. Boolean algebra, relation to switching circuits.

D. Finite state machines, state diagrams, machine homomorphism.

E. Formal languages, context-free languages, recognition by machine.

F. Groups, semigroups, monoids, permutations and sorting, representations by machines, group codes.

G. Modular arithmetic, Euclidean algorithm.

H. Optional topics: linear machines, Turing machines and related automata; Polya's enumeration theorem; finite fields, Latin squares and block design; computational complexity.

APPLIED ALGEBRA TEXTS

1. Dornhoff, Lawrence and Hohn, Frantz, *Applied Modern Algebra,* Macmillan, New York, 1978.

2. Fisher, James, *Application-Oriented Algebra,* T. Crowell Publishers, New York, 1977.

3. Johnsonbaugh, Richard, *Discrete Mathematics,* Macmillan, New York, 1984.

4. Korfhage, Robert, *Discrete Computational Structures,* Academic Press, New York, 1974.

5. Liu, C.L., *Elements of Discrete Mathematics,* McGraw Hill, New York, 1977.

6. Preparata, Franco and Yeh, Robert, *Introduction to Discrete Structures,* Addison-Wesley, Reading, Mass., 1973.

7. Prather, Robert, *Discrete Mathematical Structures for Computer Sciences,* Houghton Mifflin, Boston, 1976.

8. Stone, Harold, *Discrete Mathematical Structures and Their Applications,* Science Research Associates, Chicago, 1973.

9. Tremblay, J. and Manohar, R., *Discrete Mathematical Structures with Applications in Computer Sciences,* McGraw Hill, New York, 1975.

Numerical Analysis Course

In any elementary numerical analysis course a balance must be maintained between the theoretical and the application portion of the subject. Normally, such a course is designed for sophomore and junior students in engineering, mathematics, science, and computer science. Students should be introduced to a wide selection of numerical procedures. The emphasis should be more on demonstrations than on rigorous proofs (however, this is not meant to slight necessary theoretical aspects of error analysis). At least one or two applied problems from each of the major topics should be included so that

students have a good understanding of how the art of numerical analysis comes into play.

The course outline below presents a good selection of topics for a one-semester course. Error analysis should be continuously discussed throughout the duration of the course so as to stress the effectiveness and efficiency of the methods. Alternative methods should be contrasted and compared from the standpoint of the computational effort required to attain desired accuracy.

An optional approach to this course would emphasize a full discussion (with computer usage) of one procedure for each course topic (after the computer arithmetic introduction). A sample of five such procedures is:

1. The Dekker-Brent algorithm (see UMAP module No. 264).

2. A good linear equation solver involving LU-decomposition.

3. Cubic spline interpolation.

4. An adaptive quadrature code.

5. The Runge-Kutta-Fehlberg code RKF4 with adaptive step determination.

Weekly assignments should include some computer usage; in total, four or five computer exercises for each major topic. Students should do computer work for larger applied programs in small groups. However, the concept of utilizing "canned" programs with minor modifications should be stressed. Such an approach nicely brings out the strong interdependence between computers and numerical analysis yet does not overemphasize the efforts necessary to program a problem. An interactive computer system using video terminals is ideal for this course. Microcomputers and even handheld calculators can also be used effectively. One or two applied homework problems from each of the main topics keep students aware of the balance that is necessary between the art and the science of numerical analysis. Prerequisites for this course should be a year of calculus including some basic elementary differential equations and a computer science course.

For schools on a quarter system, two quarters should be a minimal requirement and the above material would be more than ample. One should spend the first quarter on numerical solutions of algebraic equations and systems of algebraic equations and the last quarter on the other topics.

COURSE OUTLINE

A. *Computer arithmetic.* Discretization and round-off error; nested multiplication.

B. *Solution of a single algebraic equation.* Initial discussion of convergence problems with emphasis on meaning of convergence and order of convergence;

Newton's method, Bairstow's method; interpolation.

C. *Solution systems of equations.* Elementary matrix algebra; Gaussian methods, LU decomposition, iterative methods, matrix inversion; stability of algorithms (examples of unstable algorithms), errors in conditioned numbers.

D. *Interpolating polynomials.* Lagrange interpolation to demonstrate existence and uniqueness of interpolating polynomials and for calculation of truncation error terms; splines, least squares, inverse interpolation; truncation, inherent errors and their propagation.

E. *Numerical integration.* Gaussian quadrature, method of undetermined coefficients, Romberg and Richardson extrapolation (for both integration and differentiation), Newton-Cotes formulas, interpolating polynomials, local and global error analysis.

F. *Numerical solution of ordinary differential equations.* Both initial value and boundary value problems; Euler's method, Taylor series method, Runge-Kutta, predictor-corrector methods, multistep methods; convergence and accuracy criteria; systems of equations and higher order equations.

If this course has an enrollment of under 25 students, non-standard testing can be considered, such as a take-home midterm. At the end of the term, instead of the traditional three hour examination, each student can write an expository paper exploring in greater depth one of the topics introduced in class or investigating a subject not included in the work of the course, either approach to include computational examples with analysis of errors. (Since most of the students will not have had previous experience in writing a paper, topics may be suggested by the instructor or must be approved if student devised; scheduled conferences and preliminary critical reading of papers guard against disastrous attempts or procrastination.) Some examples of final projects are: spline approximations; relaxation methods; method of undetermined coefficients in differentiation and integration; least squares approximations; parabolic (or elliptic or hyperbolic) partial differential equations; numerical methods for multi-dimensional integrals; multi-step predictor-corrector methods.

NUMERICAL ANALYSIS TEXTS

1. Cheney, Ward and Kincaid, David, *Numerical Mathematics and Computing,* Brooks/Cole, Monterey, Calif., 1980.
2. Conte, S. and DeBoor, C., *Elementary Numerical Analysis,* McGraw Hill, New York, 1978.
3. Gerald, Curtis F., *Applied Numerical Analysis, 2nd Edition,* Addison-Wesley, Reading, Mass., 1978.
4. Forsythe, G.E. and Moler, C.B., *Computer Solutions of Linear Algebraic Systems,* Prentice-Hall, Englewood Cliffs, New Jersey, 1967.
5. Hamming, R.W., *Numerical Methods for Scientists and Engineers, 2nd Edition,* McGraw Hill, New York, 1973.
6. James, M.L.; Smith, G.M.; Wolford, J.C., *Applied Numerical Methods for Digital Computation,* Harper & Row, New York, 1985.
7. Ralston, Anthony and Rabinowitz, Philip, *First Course in Numerical Analysis,* McGraw Hill, New York, 1978.

Panel Members

ALAN TUCKER, CHAIR, SUNY, Stony Brook.

RICHARD ALO, Lamar University.

WINIFRED ASPREY, Vassar College.

PETER HILTON, Case Western Reserve/Battelle Institute.

DON KREIDER, Dartmouth College.

WILLIAM LUCAS, Cornell University.

FRED ROBERTS, Rutgers University.

GAIL YOUNG, Case Western Reserve.

Calculus

This chapter contains the report of the Subpanel on Calculus of the CUPM Panel on a General Mathematical Sciences Program, reprinted with minor changes from Chapter II of the 1981 CUPM report entitled RECOMMENDATIONS FOR A GENERAL MATHEMATICAL SCIENCES PROGRAM.

Rationale

The Calculus Subpanel was charged with examining the traditional calculus sequence of the first two years of college mathematics: two semesters of single-variable calculus; one semester of linear algebra; one semester of multivariable calculus. In approaching this task, the subpanel considered syllabi through which this sequence is implemented at various colleges and universities, the syllabus for the Advanced Placement Program in Calculus, and alternatives to calculus as the entry-level course in the mathematical sciences, for example, finite mathematics or discrete methods.

The subpanel eventually came to the conclusion that the rationale for certain parts of the traditional calculus sequence remains valid, although some restructuring and increased flexibility are warranted to reflect the differing mathematical requirements of the social and biological sciences and, increasingly, of computer science. The general recommendations of the subpanel are thus:

1. To make no substantive changes in the first semester of calculus;
2. To restructure the second semester around modeling and computation, although leaving it basically a calculus course;
3. To branch to three independent courses in the second year:
   a. Applied Linear Algebra,
   b. Multivariable Calculus (in dimensions 2 and 3),
   c. Discrete Methods.

Descriptions of the first and second semesters of calculus, applied linear algebra, and multivariable calculus are given below. The discrete methods course is discussed in the first chapter, "Mathematical Sciences."

The subpanel views its recommendations as conservative. Tony Ralston has argued, for example, that calculus need not be the entry-level course in the mathematical sciences and that a course in discrete methods is a reasonable alternative, better serving some areas such as computer science (see "The Twilight of the Calculus," which appeared under the title "Computer Science, Mathematics, and the Undergraduate Curricula in Both" in the *American Mathematical Monthly*, 88:7 (1981) 472-485). In his view, to ignore discrete methods, even in the first two years of college mathematics, would be absurd in this day.

The subpanel does not disagree with the general sense of this position. On the other hand, the subpanel feels that the language, spirit, and methods of traditional calculus still permeate mathematics and the natural and social sciences. To quote Ralston himself, "The calculus is one of man's great intellectual achievements; no educated man or woman should be wholly ignorant of its elements." Perhaps the time is not far off when calculus will be displaced as the entry-level course, but it has not arrived yet.

The place for rigor. The subpanel believes strongly that, in the first two years, theorems should be *used* rather than *proved*. Certainly correct statements of theorems such as the Mean Value Theorem or l'Hôpital's Rule should be given; but motivation, as long as it is recognized as such, and usage are more important than proofs. The place for theoretical rigor is in later upper-level courses. In this regard, the subpanel agrees with the program philosophy outlined in the first chapter, "Mathematical Sciences."

First Semester Calculus

The first semester of calculus, especially, contains a consensus on essential ideas that are important for modeling dynamic events. This course has evolved through considerable effort in the mathematical community to present a unified treatment of differential and integral calculus, and it serves well both general education and professional needs. It is historically rich, is filled with significant mathematical ideas, is tempered through its demonstrably important applications, and is philosophically complete. Most syllabi for its teaching cover the usual topics:

A. *Limits and continuity.*
B. *Differentiation rules.*
C. *Meaning of the derivative.* Applications to curve sketching, maximum-minimum problems, related rates, position-velocity-acceleration problems.
D. *Antidifferentiation.*

E. *The definite integral and the Fundamental Theorem of Calculus.*

F. *Trigonometric functions.*

G. *Logarithmic and exponential functions.* Including a brief exposure to first-order, separable differential equations (with emphasis on $y' = ky$).

The first (and second) calculus courses should be 4- or 5-credit hour courses. If less time is available, topics will have to be pushed later into the calculus sequence, with some multivariate calculus material left for an analysis/advanced calculus course. Mathematics courses should not rush trying to cover unrealistic syllabi.

It might be desirable to add more non-physical sciences examples to C (e.g., a discussion of the use of the word "marginal" in economics), although serious modeling examples should be postponed to the second semester. Integration as an averaging process can be included in E, but applications and techniques (numerical or algebraic) of integration are better left to the second semester. Exponential growth and decay are important concepts that must be emphasized in G.

Second Semester Calculus

There does not appear to be much slack or fat in the first semester of calculus. It is in the second semester, therefore, when numerical techniques, models, and computer applications can be introduced. Unlike the first semester of calculus, the second semester does not enjoy the same consensus on either its central theme or its content. It tends to be a grab bag of "further calculus topics"—further techniques of integration, more applications of integration, some extension of techniques to the plane (parametric equations), sequences and infinite series, and more differential equations. Each of these topics is, in isolation, important at some stage in the training of scientists and mathematicians. But it is less clear that packaging them in this way and having them occupy this critical spot in the curriculum is justified today, given the pressing needs of computer science and the non-physical sciences.

From time to time it has been urged that multivariable calculus should be started during the second semester. But few institutions have implemented this suggestion. And the subpanel believes that, in the meantime, higher priorities for the second course have materialized in the form of applications and computing.

The subpanel considered recommending branching in the curriculum after the first semester of calculus, with students advised to take courses more directly relevant to their career goals. But it finally concluded that there

are still substantial reasons for keeping students in one "track" through the first two courses. In most American colleges, a "choice" in the second course would require most students to be thinking seriously about career goals within a few weeks of arriving on campus as freshmen. This does not strike us as realistic nor in the best interests of liberal education. Moreover, we continue to feel that many of the ideas and technical skills arising in the second calculus course are reasonable to include at this point in the curriculum. Thus, the final conclusion is that a restructuring and change of emphasis in the second semester calculus course is preferable to its replacement.

The Calculus Subpanel recommends the following changes in the second calculus course:

A. *An early introduction of numerical methods.* Implemented through simple computer programs. Solving one (or a system of two) first-order differential equation(s).

B. *Techniques of integration.* General methods such as integration by parts, use of tables, and techniques that extend the use of tables such as substitutions and (simple) partial fraction expansions; less emphasis should be placed on the codification of special substitutions.

C. *Numerical methods of integration.* Examples where numerical and "formal" methods complement each other, e.g., evaluating improper integrals where substitutions or integration-by-parts make the integral amenable to efficient numerical evaluation.

D. *Applications of integration.* Illustrate the "setting up" of integrals as Riemann sums. The emphasis should be on the modeling process rather than on "visiting" all possible applications of the definite integral.

E. *Sequences and series.* These topics should have substantially changed emphasis:

1. Sequences should be elevated to independent status, defined not only through "closed formulas" but also via recursion formulas and other iterative algorithms. Estimation of error and analysis of the rate of convergence should accompany some of the examples.

2. Series should appear as a further important example of the idea of a sequence. Power series, as a bridge from polynomials to special functions, should figure prominently. Specialized convergence tests for series of constants can be de-emphasized.

3. Approximation of functions via Taylor series, and estimation of error, accompanied by im-

plementation of such approximations on a computer.

F. *Differential equations.* Should be treated with less (but not zero) emphasis on special methods for solving first-order equations and constant coefficient linear equations (especially the non-homogeneous case). More valuable would be: vector field interpretation for first-order equations, numerical methods of solution, and power series methods for solving certain equations. Applications should arise in mathematical modeling contexts and both "closed form" and "numerical" solutions should be illustrated.

The new second course in calculus does not differ radically in content from the traditional second semester course. It is a conservative restructuring that can be taught from existing textbooks and based on modest modifications of many existing syllabi. But the intended change in "flavor" and emphasis should be more dramatic. About twelve lectures (of the usual 40 lectures) must be modified substantially to achieve the desired computer emphasis. Numerical algorithms will thus figure prominently, along with the formal techniques of calculus. Concepts not usually in a calculus course such as error estimation, truncation error, round-off error, rate of convergence, and bisection algorithms will be included. The theme for the course will be "calculus models." Consideration of even a few UMAP-type models would be enough to change the nature of the course significantly and to provide the intended "tying together" of the traditional calculus topics that are included in the course.

A syllabus for the course could be constructed by starting with the second calculus course described in the CUPM report, *A General Curriculum for Mathematics in Colleges* (revised 1972), or with the Advanced Placement BC Calculus Syllabus. Topics to be diminished or omitted include: emphasis on special substitutions in integrals, l'Hôpital's rule except as it arises naturally in connection with Taylor series, polar coordinates, vector methods, complex numbers, non-homogeneous differential equations and the general treatment of constant-coefficient homogeneous linear differential equations. Many of these topics will appear in examples but will not be emphasized in themselves.

Intermediate Mathematics Courses

Although the Calculus Subpanel recommends retaining a single track for students during their first year, it just as strongly recommends that three different courses be available from which students choose (with advising) their intermediate mathematics courses. Two of these courses, whose descriptions follow, are Applied Linear Algebra and Multivariable Calculus. The third, Discrete Mathematics, is described in the first chapter, "Mathematical Sciences."

Applied Linear Algebra

For a large part of modern applied mathematics, linear algebra is at least as fundamental as calculus. It is the prerequisite for linear programming and operations research, for statistics, for mathematical economics and Leontief theory, for systems theory, for eigenvalue problems and matrix methods in structures, and for all of numerical analysis, including the solution of differential equations. The attractive aspect about these applications is that they make direct use of what can be taught in a semester of linear algebra. The course can have a sense of purpose, and the examples can reinforce this purpose while they illustrate the theory.

A number of major texts have arrived at a reasonable consensus for a course outline. Their outlines are well matched with the needs of both theory and application. Applications can include such topics as systems of linear differential equations, projections and least squares. But the subpanel strongly recommends that more substantial applications to linear models should be a central part of the construction of the course. Many different applications of this kind are accessible and can be found in the texts mentioned. Thus, no rigid outline is required. The development of the subject moves naturally from dimension 2 to 3 to n, and although that is an easy and familiar step, it nevertheless represents mathematics at its best. The combination of importance and simplicity is almost unique to linear algebra. Linear programming is an excellent final topic in the course. It brings the theory and applications together.

The changes in this course are ones of emphasis that recognize that the course must be more than an introduction to abstract algebra. Abstraction remains a valuable purpose, and linearity permits more success with proofs than the epsilon-delta arguments of calculus. However, the main goal is to emphasize applications and computational methods, opening the course to the large group of students who need to *use* linear algebra.

TEXTS

1. Hill, Richard, *Elementary Linear Algebra*, Academic Press, New York, 1986.
2. Kolman, Bernard, *Introductory Linear Algebra with Applications*, Macmillan, New York, 1979.

3. Rorres, Chris and Anton, Howard, *Applications of Linear Algebra,* John Wiley & Sons, New York, 1979 (paperback supplementary text).
4. Strang, Gil, *Linear Algebra and Its Applications, 3rd Edition,* Harcourt Brace Jovanovich, San Diego, 1988.
5. Tucker, Alan, *A Unified Introduction to Linear Algebra,* Macmillan, New York, 1988.
6. Williams, Gareth, *Linear Algebra with Applications,* Allyn and Bacon, Boston, 1984.

Multivariable Calculus

This is the traditional multivariable calculus course at many colleges and universities. It is not a new course, but for many schools it would represent a movement in the direction of "concrete" treatment of multivariable calculus rather than the more recent elegant treatments making heavy use of linear transformations and couched in general (high dimensional) terms. The course begins with an introduction to vectors and matrix algebra. Topics include Euclidean geometry, linear equations, and determinants. The remainder of the course is an introduction to multivariable calculus, including the analytic geometry of functions of several variables, definitions of limits and partial derivatives, multiple and iterated integrals, non-rectangular coordinates, change of variables, line integrals, and Green's theorem in the plane.

Differential Equations

The Calculus Subpanel has considered the place of differential equations in the curriculum. It recommends that the topic be treated at two levels:

1. Through methods and examples involving differential equations, spiraled through the calculus sequence, and
2. Through a substantial course in differential equations, available to students upon completion of the first-year calculus sequence and applied linear algebra.

We note here topics in differential equations that are part of the preceding courses:

- Solutions of $y' = ky$ occur in the first semester of calculus. Exponential growth and decay are discussed.
- Solution of second order linear differential equations are included in the second semester of calculus. Oscillating solutions occur as examples. In addition, geometrical interpretations (direction field), numerical solutions and power series solutions are included.

- Applied Linear Algebra includes the solution of linear constant coefficient systems of differential equations using eigenvalue methods.

Although the Calculus Subpanel has *not* recommended a full course in differential equations in the calculus sequence of the first two years, it has suggestions for a subsequent course. Such a course should not be a compendium of techniques for solving in closed form various kinds of differential equations. Libraries are full of cookbooks; one hardly needs a course to use them. What is important is to develop carefully the models from which differential equations spring. Modeling obviously means more than an application such as:

> According to physics, the displacement $x(t)$ of a weight attached to a spring satisfies $x'' - bx' + kx = 0$. Solve for $x(t)$ given that $b = 2, k = 3, x(0) = 1, x'(0) = 0$.

For a more serious approach to applications, we refer to the art forgery problem at the beginning of Braun (see below) or indeed almost any of the models discussed in the suggested texts.

The meaning of the word "solution" must be scrutinized. Different viewpoints must be introduced—numerical, geometric, qualitative, linear algebraic and discrete.

A possible syllabus for a differential equations course is:

A. *First-order equations.* Models; exact equations; existence and uniqueness and Picard iteration; numerical methods.
B. *Higher-order linear equations.* Models; the linear algebra of the solution set; constant coefficient homogeneous and non-homogeneous; initial value problems and the Laplace transform; series solutions.
C. *Systems of equations and qualitative analysis.* Models; the linear algebra of linear systems and their solutions; existence and uniqueness; phase plane; nonlinear systems; stability.

Since some of these topics will have already been introduced in courses from the calculus sequence, there may be time for a brief discussion of partial differential equations and Fourier series. Existence and uniqueness theorems are included here only because of the light they or their proofs might shed on methods of solution (e.g., Picard iteration).

TEXTS

The course can be taught using any of the many reasonable differential equations texts with a modest amount of applications, supplemented by:

Braun, Martin, *Differential Equations and Their Applications, Second Edition,* Springer-Verlag, New

York, 1978.

Braun remains the only text to build extensively on applications, but it has the serious drawback that it is based on single-variable calculus and avoids linear algebra.

A somewhat radical alternative is a theoretical course involving more qualitative or topological analysis emphasizing systems of equations. The subpanel does not suggest a syllabus, but refers instead to V.I. Arnold, *Ordinary Differential Equations*, MIT Press, Cambridge (paperback), 1978.

This course would have applied linear algebra and multivariable calculus as prerequisites and could be taken as early as the second semester of the sophomore year if the two prerequisites were taken concurrently the previous semester.

Subpanel Members

DON KREIDER, CHAIR, Dartmouth College.
ROSS FINNEY, Educational Development Center.
JOHN KENELLY, Clemson University.
GIL STRANG, MIT.
TOM TUCKER, Colgate University.

Core Mathematics

This chapter contains the report of the Subpanel on Core Mathematics of the CUPM Panel on a General Mathematical Sciences Program, reprinted with minor changes from Chapter III of the 1981 CUPM report entitled RECOMMENDATIONS FOR A GENERAL MATHEMATICAL SCIENCES PROGRAM.

New Roles for Core Mathematics

In the 1960's CUPM extensively examined curriculum in core mathematics—upper division subjects that comprise the trunk from which the other specialized branches and applications of mathematics emerge. It reviewed and revised its recommendations in 1972. See the *Compendium of CUPM Recommendations* published by the Mathematical Association of America, especially the 1972 revision of the *General Curriculum for Mathematics in Colleges*. With the current restructuring of the mathematics major into a mathematical sciences major, new questions have been raised about core mathematics curriculum. These questions concern the role of core mathematics in a mathematical sciences major as much as syllabi of individual courses. This chapter focuses on four questions that were addressed to the Core Mathematics Subpanel by the parent Panel on a General Mathematical Sciences Program.

The members of this subpanel represent a variety of institutions, public and private, liberal-arts colleges, and research-oriented universities. All the members have seen at their institutions a divergence of the mathematics major from its form during their own undergraduate training, as career opportunities for mathematics majors have changed. In part, the members lament the passing of the mathematics program that nurtured their love of mathematics. At the same time they acknowledge the challenge of the diversity of the present and future. They realize that it is not now realistic for CUPM to recommend a core set of pure mathematics courses to be taken by all mathematical sciences majors in every institution.

While the mathematics major has generally broadened towards a mathematical sciences major, it is still possible for an institution, large or small, to elect to retain a traditional pure mathematics major, alone or in conjunction with an applied mathematics major. But it is clearly more appropriate to work within current realities to fashion a unified mathematical sciences major with diminished pure content, a major incorporating both breadth and selective depth. (If size warrants, the unified major can have several tracks, one for preparation for graduate study in mathematics.) This subpanel is concerned with the role in a mathematical sciences major of upper-level core mathematics courses, and more generally with appreciation of the depth and power of mathematics.

A prime attribute of a person educated in mathematical sciences is his or her ability to respond when confronted with a mathematical problem, whether in pure mathematics, applied mathematics, or one which uses mathematics that the person has not seen before. Our students should be prepared to function as professionals in areas needing mathematics not by having learned stock routines for stock classes of problems but by having developed their ability in problem solving, modeling and creativity. This general pedagogical theme, that was stressed throughout the first chapter "Mathematical Sciences," guided the thinking of the Core Mathematics Subpanel.

The report of the Core Mathematics Subpanel is meant to be supportive rather than directive. What an individual department does should reflect its constituency of students, their needs, their numbers, and the goals, character and size of the institution.

Four Questions

QUESTION 1: *Is there a minimal set of upper-level core mathematics (algebra, analysis, topology, geometry) that every mathematical sciences major should study?*

ANSWER: No. There is no longer a common body of pure mathematical information that every student should know. Rather, a department's program must be tailored according to its perception of its role and the needs of its students. Whether pure mathematics is required of all in some substantial way; whether it is used as an introduction to advanced work of applied nature or as a completion to an applied program; or whether pure mathematics is simply one track in a collection of programs in a large department will be an institutional option. Departments must recognize this fact, establish their programs with a clear understanding of objectives that are being met, and be prepared to share and explain these perceptions with their students. The limited resources of smaller departments must be exploited with great efficiency and wisdom. Such departments may face a difficult decision of whether to abandon certain traditional branches of mathematics entirely in order to offer courses and tracks best suited to their students.

The underlying problem is that students enter college with much less mathematics than they used to, but they expect to leave with more. There is a wide span of preparation among entering college students, they want an education that is specific to chosen career goals, and the levels of mathematical and computational skills and sophistication that accompany these goals have risen. Core courses such as abstract algebra and analysis are valuable for continuing study in many fields, but they are not essential for all careers.

The Core Mathematics Subpanel and the parent Mathematical Sciences Panel jointly recommend that all mathematical sciences students take a sequence of two courses leading to the study of some subject in depth (see the first chapter, "Mathematical Sciences").

QUESTION 2: *Should there be major changes in the content or mode of instruction of upper-level core mathematics courses?*

ANSWER: While there will continue to be some students who plan to move toward a doctorate in pure, or applied, mathematics and an academic career, the mathematical sciences major is seen by most students as preparation for immediate employment or for Masters-level graduate training in areas outside of mathematics (but where mathematical tools are needed). Thus mathematics departments can no longer view their upper-division courses as a collection of courses that faculty wish they had had prior to admission to graduate school. Rather, departments must offer pure mathematics courses that are compatible with the overall goals of a mathematical sciences major, courses that are intellectually and pedagogically complete in themselves, courses that are both the beginning and the end of most students' study of the subject. The main objective in such courses now is developing a deeper sense of mathematical analysis and associated abstract problem-solving abilities. In these courses students learn how to learn mathematics.

There is always a continuing need to re-examine the nature and content of any course. Some courses carry baggage that may be there largely for historical reasons. A frequent example of this is the traditional course in differential equations which is populated by isolated discoveries of the Bernoulli clan (and lacking in discussion of numerical methods). Instructors are slow to discard topics that have a strong aesthetic appeal (for the instructors) but are no longer important building blocks

in the field. Syllabi and approaches in pure mathematics courses must be adapted to changing constituencies with a careful balance of learning new concepts and modes of reasoning and of using these constructs, a balance of "listening" and "doing." Students should emerge from a course feeling that they have become junior experts in some topics: they should know facts and relationships, know some of the "whys" behind this mathematics.

It would be desirable for courses to be structured with review stages that require reflection by students of what analysis to use to solve a problem. The courses need to contain assignments that ask for short proofs of results and for application of concepts and techniques from one problem to another (apparently unrelated) problem. Proper judgment in the selection of a method of analysis is the key both to constructing mathematical proofs and to problem solving in applied mathematics; nurturing this ability is the critical challenge to instructors. Students should be required to present material both orally and in writing on a regular basis. Since students do not have to know a standard body of theorems for graduate study, the course content in algebra, analysis, topology and geometry can vary according to faculty interests and possible ties with stronger quantitative areas of an institution (e.g., physics or biology).

The density of proofs in an upper-level course is always a controversial issue. It is traditional to feel that one objective of such a course is to teach students how to construct proofs. However, this skill comes slowly and seldom arouses the same pleasure in students as it does in instructors. Some proofs are needed in any upper-level mathematics course to knit together the entire structure that is being presented, but one should probably aim at piecewise rigor rather than a Landauesque totality. Students' mathematical maturity will develop as much, and it will be far less painful.

The preceding pedagogical goals in core mathematics must accommodate the reality that courses such as abstract algebra may only be offered in alternate years and that two-semester sequences or courses with core mathematics prerequisites will be difficult to schedule. With a broad mixture of students in infrequently-offered courses, instructors must be sensitive to the discouragement some students may feel in the presence of more sophisticated seniors.

QUESTION 3: *How can the full scope of mathematics be conveyed to students? Should this be done by one-semester survey courses that cover a range of fields?*

ANSWER: Students pursuing specific career goals in mathematical sciences and those taking upper-level mathematical "service" courses need to be made aware of the depth and breadth of mathematics and the greater mathematical maturity that their subsequent careers may demand. Mathematical survey courses do not appear to be the answer. They will not be able to move beyond vocabulary and notation to give any sense of global structure in any of the fields covered.

Physicists seem to have been remarkably successful in communicating some understanding about the "big picture" to their students and laymen through expository articles that treat highly technical subjects by presenting only a projection or shadow of the true structure, but doing so in a way that does not seem to offend their consciences. Similar approaches should be possible in mathematics using expository *American Mathematical Monthly*, *Mathematics Magazine*, or *Scientific American* articles. Following the reading of such an article, a (once-a-week) class would discuss concepts, technicalities and applications in the article plus additional examples. Natural topic areas are complex analysis and two-dimensional hydrodynamics; number theory and public key cryptography; calculus of variations and soap films; queueing theory and, say, toll road design.

More traditional ways of projecting the wide-ranging nature of mathematics are by rotation of courses and by providing seminars, extracurricular mathematical activities, summer work opportunities, and by references and linkages to mathematics in courses in other departments. This breadth should also give a sense of the rapidly changing nature of uses of mathematics and of the need of learning how to learn mathematics.

QUESTION 4: *Should pure mathematics courses be postponed for most students until the senior year to follow and abstract from more applied courses earlier in the curriculum?*

ANSWER: Many mathematical sciences students who prefer problem solving to theory appear to have considerable difficulty in their sophomore or junior years with abstract core mathematics. For these students, core mathematics may better wait until a senior year "capstone" course(s) that builds on maturity developed in earlier problem solving courses. This course (preferably year-long if only one such course is required) in a subject such as analysis or abstract algebra would build a student's capacity (and appetite) for abstraction and proof and for solving complex problems involving a combination of analytical techniques. The course would seek depth rather than breadth. The course should link abstract concepts with their concrete uses in previous courses, such as integration concepts used in limiting probability distributions. It should illustrate in several

ways the power and usefulness of mathematical abstraction and generalization.

There are two important provisos about senior-year courses. First, when core courses cannot be offered every year, they obviously must be accessible to most juniors. Second, the mathematically gifted student (whether a mathematics major or not) must be able to take such senior core courses in the sophomore year without needing applied prerequisites that other students naturally take before the core course. Such gifted students today are often directed towards popular careers such as engineering or medicine and by their senior year would be too immersed in professional training to take the pure mathematics course that would reveal their mathematical research potential.

It is worthwhile recalling that before 1950 few colleges offered regular courses in abstract algebra, topology, or up-to-date advanced calculus. The 1950's and 1960's were memorable in mathematics education, but today's students must be viewed as in the historical mainstream rather than as slow in learning to handle abstractions.

Individual institutions will differ greatly in the design of such senior courses. As noted in the discussion of Question 2, these courses should require oral and written student presentations. The spirit of this recommendation could be achieved with a year-long course in a subject such as differential equations or combinatorics that begins with applications and leads to abstraction or a course that begins with abstraction and leads to applications.

Sample Course Outlines

In this section we discuss two approaches to the fundamental upper-level core subjects of abstract algebra and analysis. We suggest an ideal treatment and then a more modest version that is appropriate for most current mathematical sciences students. The descriptions are stated in terms of student objectives.

The philosophy behind each of the course descriptions is that the student needs a working understanding of the subject far more than a detailed intensive and critical knowledge. The instructor's central goal is to teach the student how to learn mathematics, expecting that students will correctly retain only a tiny portion of what was taught, but that when they need to refresh their knowledge, they will be far better able to do so than if they had never taken the course. Proofs are not of major importance, but in both approaches students should be able to understand what the hypotheses of a theorem mean and how to check them. They should also be able to detect when seemingly plausible statements are false (and should be shown counterexamples to such statements; e.g., integrals that should converge but do not).

Abstract Algebra I (Ideal)

A. Give the student a guided tour through the algebraic "zoo," so that he or she knows what it means to be a group, a ring, a field, an associative algebra, etc. Include associated concepts such as category, morphism, isomorphism, coset, ideal, etc.

B. Show the student useful ways for generating one algebraic structure out of another, such as automorphism groups, quotient groups, algebras of transformations, etc.

C. Give the student an understanding of the basic structure theorems for each of the algebraic systems discussed, as well as an understanding of their proofs.

D. Give the student experience in using the preceding ideas and constructions and seeing how these ideas arise in other branches of mathematics (analysis, number theory, geometry, etc.).

E. Show the student how algebra is used in fields outside of mathematics, such as physics, genetics, information theory, etc.

Abstract Algebra II (Modest)

A. Combine parts of A and B of Course I by showing students at least two different types of algebraic structures and several instances in which such an algebraic structure evolved or is constructed out of another mathematical structure. The goal is for a student to be able to recognize when a situation has aspects that lend themselves to an algebraic formulation; e.g., rings out of polynomials.

B. Describe part of the theory for one of the structures introduced in A and illustrate several of the deductive steps in the theory. Students should see the nature of tight logical reasoning and the usefulness of algebraic concepts, as well as come to appreciate the cleverness of the theory's discoverers.

C. Discuss at least one application of algebra outside of mathematics.

D. Assign students a variety of problems which require recognition of algebraic structures in unfamiliar forms, proof of small deductive steps, and use of theory in B.

Analysis I (Ideal)

A. Give the student a working knowledge of point set topology in R^n and analogous concepts for a metric space.

B. Study the class of continuous maps from a region in R^n into R^m, and the special properties of maps in class C' and C''.

C. Study integration of continuous and piecewise continuous functions over appropriately chosen sets, bounded and unbounded, and then extend this to integration with respect to set functions.

D. Extend to the theory of differential forms and develop a relationship between differentiation of forms and the boundary operator, via Stokes' theorem.

Analysis II (Modest)

A. Give the student a glossary of terms in point set topology, appropriate also to a metric space and applied to R^n, and practice in their meanings. (Do not prove inter-relations, but state them clearly.)

B. Introduce the class of C'' maps from R^n into R^m, and discuss a few problems involving such functions, each motivated by a concrete "real" situation. Solve each of the problems by stating and illustrating the appropriate general theorems, and in a few cases, sketching part of the proofs.

C. Discuss integration in terms of measurement and averaging, extend this to R^n, and explain briefly techniques of numerical integration. At all stages give attention to improper integrals.

D. Extend the notion of function to differential forms, illustrated with physical and geometric examples. Motivate Stokes' theorem as the analogue of the fundamental theorem of calculus, and arrive at a correct formulation of it without proof. Illustrate the theorem with examples, including some involving the geometric topology of surfaces; if students' background is appropriate, examples in physics (hydrodynamics or electromagnetism) should be given.

An analysis course can also be given an "advanced calculus" emphasis including topics such as Fourier series and transforms, special functions, and fixed-point theorems, with applications of these topics to differential equations. For further discussion of this approach, see versions one and three of Mathematics 5 in the CUPM recommendations for a *General Curriculum in Mathematics for Colleges* (revised 1972).

Subpanel Members

PAUL CAMPBELL, CHAIR, Beloit College.
LIDA BARRETT, Northern Illinois University.
R. CREIGHTON BUCK, University of Wisconsin.
MARGARET HUTCHINSON, University of St. Thomas.

Computer Science

This chapter contains the report of the Subpanel on Computer Science of the CUPM Panel on a General Mathematical Sciences Program, reprinted with minor changes from Chapter IV of the 1981 CUPM report entitled RECOMMENDATIONS FOR A GENERAL MATHEMATICAL SCIENCES PROGRAM.

A Growing Discipline

Computer Science is a new and rapidly growing scientific discipline. It is distinct from Mathematics and Electrical Engineering. The subject was once closely identified in mathematicians' minds with writing computer programs. In the beginning, however, computer scientists concentrated on the discipline's mathematical theories of numerical analysis, automata, and recursive functions, as well as on programming. In the past decade, theories developed to understand problems in software design (compilers, operation systems, structured programs, etc.) have blossomed. These theories involve the analysis of complex finite structures, and in this sense have a strong mathematical bond with the finite structures common in operations research and diverse areas of applied mathematics.

More importantly, these computer science theories are needed by analysts who design algorithms for complex problems in the mathematical sciences. For this reason, all mathematical sciences students must be given an introduction to the basic concepts of computer science. Further, facility in computer programming is required of all mathematical sciences students so that they can perform practical computations in mathematical sciences courses and in subsequent mathematical sciences careers.

Although only one-third of the country's colleges and universities now have computer science departments, the number of students currently majoring in computer science taught in a computer science department (approximately 50,000 students) is greater than the number of all majors in mathematics, mathematical sciences, and applied mathematics. The computer science recommendations in this chapter are designed for institutions where computer science is taught in a mathematical sciences department or in a mathematics department. When a separate computer science department exists, that department's diversity of computer science offerings will enhance a mathematical sciences major. A mathematical sciences undergraduate program and a computer science undergraduate program should complement one another to the advantage of both departments and their students (for example, see the description of the interaction at Potsdam State in Chapter I, "A General Mathematical Science Program").

Introductory Courses

The foundation for a computer science component in a mathematics department is a one-year introductory sequence. Courses CS1 and CS2, proposed in the Association of Computing Machinery Curriculum 78 (see last section of this chapter), are excellent models for this year sequence. The Subpanel on Computer Science endorses the objectives of these two courses, and recommends that all mathematical sciences majors should be required to take the first course and strongly encouraged to take the second course in this sequence. If the second course is not required, substantial use of computers should be an integral part of other mathematical sciences courses.

The primary emphasis in the first course should be on:

- Problem solving methods and algorithmic design and analysis,
- Implementing problem solutions in a widely used higher-level programming language,
- Techniques of good programming style, and
- Proper documentation.

Lectures should include brief surveys of the history of computing, hardware and architecture, and operating systems.

The second course should include at least one major project. The course should cover topics such as recursive programming, pointers, stacks, queues, linked lists, string processing, searching and sorting techniques. The concepts of data abstraction and algorithmic complexity should be introduced. Proofs of correctness may also be discussed.

Good design and style in programming should be emphasized throughout both courses: the use of identifiers to indicate scope, modularity, appropriate choice of identifiers, good error recovery procedures, checks for integrity of input, and appropriate commentary and documentation. Of course, efficient algorithms and coding should also be stressed. There is a strong tendency among students to worry only about whether their programs run correctly. Through class lectures and careful grading of programming assignments, the instructor must teach the students the importance of good design, style, and efficiency in programming.

A source of useful commentary about introductory computer science courses is the *SIGCSE (Special Interest Group on Computer Science Education) Bulletin*. The bulletin is published quarterly, and issue #1 each year, which contains papers presented at the SIGCSE annual meeting, is especially valuable.

Most introductory texts have many sample projects. In addition, the following three texts are good general sources of computer projects.

1. Bennett, William R., *Scientific and Engineering Problem-Solving with the Computer,* Prentice-Hall, Englewood Cliffs, New Jersey, 1976.
2. Gruenberger, Fred and Jaffray, G., *Problems for Computer Solution,* John Wiley & Sons, New York, 1965.
3. Wetherall, Charles, *Etudes for Programmers,* Prentice-Hall, Englewood Cliffs, New Jersey, 1978.

Mathematicians teaching introductory computer science often emphasize numerical computation in programming assignments. At the introductory level, the computer science issues involved in numerical computation are quite simple. Assignments requiring symbolic manipulation and data organization present more substantive programming problems and, in general, require more thought. The following is a sample assignment that could be given late in the first course:

> Write a program which obtains a five-card poker hand from some source (terminal, input deck, or file), prints the hand in a reasonably well-formatted style, and determines whether or not the hand contains a pair, three of a kind, a straight, a full house, etc.

Intermediate Courses

Intermediate-level computer science courses building on CS1 and CS2 should address basic underlying issues in computer science. In describing computer science in the first two years, the ACM Curriculum 78 report states that the student should be given "a thorough grounding in the implementation of algorithms in programming languages which operate on data structures in the environment of hardware." Thus these courses should develop general topics about algorithms, concepts in programming languages, data structures, and computer hardware.

The intermediate-level courses should be taught by a computer scientist, that is, by an individual who has significant graduate-level training in computer science (see below).

The Subpanel on Computer Science, in concurrence with ACM curriculum groups, strongly rejects the idea of a set of courses that each address a specific programming language, e.g., a sequence of advanced FORTRAN, COBOL, RPG, and APL. The argument for such a sequence is usually based on the employability of students completing it. If indeed this argument is valid, and there is some question about that, it is a short range benefit. Students completing such a sequence will soon find that the lack of underlying concepts will put them at a severe disadvantage. However, it may be acceptable, resources permitting, to have one "vocational" elective course that studies a second higher-level language such as COBOL. Of course, it is also natural to discuss new programming languages in several intermediate (and advanced) computer science courses. However, the new language would not be the focus of the course, but rather a tool used in learning and illustrating fundamental concepts.

The role of numerical and computational mathematics in computer science has diminished in recent years. While the ACM Curriculum 68 treated numerical analysis as part of core computer science, today numerical mathematics is considered by most computer scientists to be simply another mathematical sciences field that has overlap with computer science. Numerical mathematics is very important in a mathematical sciences major, but it is not a part of the computer science component.

Following the CS1 and CS2 courses, the ACM Curriculum 78 specifies six additional courses in core computer science.

CS3 Introduction to Computer Systems
CS4 Introduction to Computer Organization
CS5 Introduction to File Processing
CS6 Operating Systems and Computer Architecture
CS7 Data Structures and Algorithm Analysis
CS8 Organization of Programming Languages

The syllabi of these courses are given at the end of this chapter. Ideally, all six of these courses would be offered. A concentration or a minor in computer science would commonly consist of CS1 and CS2, followed by two of CS3, CS4, and CS5, and two of CS6, CS7, and CS8. For the purposes of a mathematical sciences program, it may be justified to place more emphasis on the software oriented areas. This would imply, if there was difficulty in offering all six courses, that CS3, CS5, CS7, and CS8 would be most useful. Then CS3, CS5, CS7, and CS8 would be offered once a year, and CS4 and CS6 offered as topics courses every other year.

At many schools, it may not be feasible to offer at least four of these intermediate courses in computer science on a regular basis. Then one can combine parts of these intermediate courses to provide a significant offering in two courses above CS1 and CS2. In this case, only two computer science courses, one elementary and one intermediate, would be offered each semester. One approach would be to combine topics from CS5 and CS7 into one course, and topics from CS3, CS4, and CS6 into the other. This would yield two courses with the following sort of syllabi (for more details about these topics, see the ACM Curriculum 78 syllabi at the end of this chapter):

A1. Algorithms for Data Manipulation
  1. Algorithm design and development illustrated in areas of sorting and research (25%)
  2. Data structure implementation (30%)
  3. Access methods (25%)
  4. Systems design (15%)
  5. Exams (5%)
A2. Computer Structures
  1. Basic logic design (15%)
  2. Number representation and arithmetic (10%)
  3. Assembly systems (35%)
  4. Program segmentation and linkage (15%)
  5. Memory management (10%)
  6. Computer systems structure (10%)
  7. Exams (5%)

This approach focuses on data structures and software issues that relate to operating systems. An alternative approach could concentrate on programming languages and algorithms involved in computer systems performance. This theme could be realized by combining topics in CS3, CS5, and CS8 into one course, and topics in CS4, CS6, and CS7 into the other course. This would yield two courses with the following syllabi:

B1. Language Types and Structures
  1. Assembly systems (25%)
  2. Program segmentation and linkage (15%)
  3. Language definition structure (10%)
  4. Data types and structures (15%)
  5. Control structures and data flow (20%)
  6. Access methods (10%)
  7. Exams (5%)
B2. Algorithms for Computer Systems
  1. Basic logic design (20%)
  2. Algorithm design and analysis (20%)
  3. Procedure activation algorithms (15%)

  4. Memory management (15%)
  5. Process management (15%)
  6. Systems design (10%)
  7. Exams (5%)

It is important to note that an individual wishing to go on from these courses to advanced work in computer science may have to make up, as deficiencies, areas in core computer science that are not represented in these condensed pairs of courses.

Concentrations and Minors

A computer science concentration in a college mathematics department can be defined as an option within a mathematical sciences major or as a "stand-alone" minor. A computer science minor should consist of about six courses, ACM Curriculum 78 courses CS1 and CS2 plus four intermediate courses.

A computer science concentration within a mathematical sciences major has three components:
  A. Mathematics: 5-plus courses;
  B. Computer Science: 4-6 courses;
  C. Applied Mathematics: 3-plus courses.

A. The mathematics component would include the three semester freshman-sophomore "calculus sequence" plus linear algebra. As recommended in Chapter I, "A General Mathematical Sciences Program," any mathematical sciences major should contain upper-level course work of a theoretical nature, typically algebra or advanced calculus. In a major with a computer science concentration, algebra is the natural area. Specifically, the applied algebra course given in Chaper I would be excellent for the computer science concentration. The course's syllabus incorporates most of the topics of the ACM 78 discrete mathematics course (required of computer science majors). A small department could offer applied algebra and standard abstract algebra courses in alternate years. Logic and automata theory are attractive electives in the mathematics component if a mathematics department wishes to focus on more theoretical aspects of computer science.

It should be noted that several computer science educators have questioned the reliance on calculus as the basic mathematics for future computer scientists; ACM Curriculum 78, for instance, requires a (freshman) year of calculus. They advocate a mathematics component based on discrete mathematics with only one semester of calculus (taught, say, in the junior year). See A. Ralston and M. Shaw, "Curriculum 78—Is Computer Science Really that Unmathematical?", *Communications ACM* 23 (1980), pp. 67-70.

B. The computer science component would include ACM Curriculum 78 courses CS1 and CS2 plus two to four intermediate courses, as described in the preceding section. The syllabi of ACM Curriculum 78 core courses are given at the end of this chapter.

C. The applied mathematics component should include a course in numerical analysis and a course in probability and statistics. The third applied mathematics course would be discrete methods, which would cover the combinatorial material in the ACM Curriculum 78 discrete mathematics course in greater depth, including operations-research-related graph modeling (see Chapter I for a full description of this course). The CUPM Mathematical Sciences Program panel recommends that all mathematics departments should offer a discrete methods course. Other good courses for the applied mathematics component are ordinarily differential equations, mathematical modeling, and operations research. The 1971 CUPM *Report on Computational Mathematics* describes courses in computational models, in combinatorial computation, and in differential equations with numerical methods; these courses combine topics from a variety of mathematical sciences and computer science courses and hence are particularly attractive to small departments.

In either the computer science concentration or minor, all six computer science courses are needed for future graduate study in computer science. Incoming graduate students with less preparation are commonly required to make up undergraduate course deficiencies.

Faculty Training

For the foreseeable future, the dominant factor affecting computer science instruction at all institutions, but particularly at smaller colleges and universities, will be the extreme shortage of qualified computer scientists in academe. At smaller colleges and universities it may therefore be effectively impossible to hire a computer scientist to teach core computer science courses. Among the possible solutions to this problem are:

1. Using adjunct faculty to teach computer science courses.
2. Using existing (non-computer science) faculty to teach computer science courses.

The first solution is acceptable for some courses. Although one cannot build a program with adjunct faculty and although staffing courses with adjunct faculty is never as desirable as using full-time faculty (e.g., student advising is a particular problem), this is a feasible way to get computer science courses taught when such faculty exist in the local community. However,

since so many smaller colleges are located away from the metropolitan areas where most technical and scientific employers of such adjunct faculty are found, this solution will not be useful to most smaller institutions.

A crucial point that must be emphasized when using existing non-computer science faculty (i.e., mathematicians) to teach computer science courses is that computer science cannot be treated like most other new mathematics course topics which mathematicians will (quickly) learn as they teach it. Mathematicians untrained in computer science are very likely to teach computer science badly, hurting both the students and the mathematics department's reputation. Therefore, if a current mathematics faculty member is to be used to teach computer science, especially beyond the first course, he or she must first acquire some formal education in computer science.

The most plausible approach to such computer science training is through some program of released time. The pertinent questions about the training are: how long? where? and how financed?

Assuming that the mathematician who is to be trained is, at most, familiar with programming in a high-level language, then full-time study for one year is the minimum period needed to acquire the background, knowledge, and experience necessary to teach several of the intermediate-level core computer science courses. Since one year is also the maximum period which would be administratively or financially feasible, this should be viewed as the canonical period for faculty training in computer science. Part-time study over a longer period or a succession of summers can also be considered. However, both because the needs to train faculty in computer science are pressing and because intermittent study is almost always less effective than continuous study, at least one faculty member in a mathematics department should have completed a one-year program of full-time study in computer science.

The most logical place at which to study computer science for the purpose of becoming able to teach it is at a university with undergraduate and graduate (preferably Ph.D.) programs in computer science. Although there are exceptions, the current level of computer science instruction in American colleges and universities is so uneven that only at such institutions can one be reasonably assured of an atmosphere in which there will be the necessary broad understanding of the principles of computer science. Such an atmosphere is particularly important for an academic mathematician preparing to teach the subject.

Another possibility which should be mentioned is for the faculty member to spend one year at one of those (relatively few) major industrial firms with good in-house training programs in computer science. An additional attraction to this idea is that it might be possible to arrange an exchange in which a member of the firm taught at the college for a year.

Methods of financing such a program of faculty training in computer science are fairly obvious:

1. Through released time at full pay from the mathematician's home institution.
2. Through grants from current, and hopefully new, federal programs; officials of both the MAA and ACM are currently pressing NSF to provide more funds for this purpose.
3. Through grants from private foundations; individual institutions and departments may be more effective than professional associations in obtaining such private funds.
4. Through corporate sponsorship of participation in in-house training programs or academic-corporate exchanges.

Computer Facilities

Facilities to support computing in mathematical sciences instruction can be provided in a variety of ways, ranging from one large centrally administered system to many small personal computing devices. The suitability of a particular means depends not only upon its intended applications, but also upon factors such as cost, ease of use, and local politics. At present, computing services in most colleges and universities are provided by a large centralized facility, the Computing Center. Growing numbers of institutions, however, are beginning to decentralize computing on campus. Three current modes of providing service are discussed below:

- Centralized facilities
- Departmental computers
- Personal computers.

There is a fourth mode that is primarily a form of access to centralized or departmental computers:

- Terminals

The second half of this section discusses the cost and ease of implementation of various applications with different types of computing facilities.

It should be noted that it is possible for an institution to form a consortium with nearby schools to operate a common central computing facility or to buy time (and services) from commercial computing centers. This option allows an institution to have a mix of computing, using large computers for problems requiring great speed or memory size, such as "number crunching," and smaller computers for student programs and other instructional purposes.

CENTRALIZED FACILITIES

Historically, so-called "economies of scale" encouraged the development of increasingly larger computers; and of increasingly larger organizations to administer them. Such computer systems are capable of providing a great variety of services with a low cost for each service. In addition, the organizations which administer these systems can play an important role in developing and supporting instructional uses of computing on campus.

On the other hand, the very size of such facilities and the organizations that administer them create certain problems. First, large systems have a high unit cost, in the range of half a million to several million dollars; replacing or enhancing such a system involves a major administrative decision. Second, instructional users of such systems must often compete with other powerful and better-financed constituencies; either separate facilities are needed to reduce competition among instructional, research, and administrative uses of the computer, or policies are needed to allocate the services provided by a single facility. And third, large organizations can be bureaucratic and inflexible.

DEPARTMENTAL COMPUTERS

For the last ten years minicomputers have provided an alternative to a large centralized facility. Lower unit costs (around $100,000 or less) and the possibility of local control have made it attractive for academic and administrative departments to acquire facilities of their own. Such facilities can be tailored to a department's needs and can provide almost as many services as a large centralized system.

Minicomputers, however, are not necessarily the answer to every department's computing needs. First, there is the question of which services they will provide. Second, there are hidden costs associated with administering any computer facility: personnel are needed to operate and maintain the facility and to provide technical assistance to users. Small departments run the risk of diverting attention from their primary task of teaching mathematics to the subsidiary task of managing such an enterprise. One way to deal with such hidden costs is for departments to contract with a central campus organization to manage their facilities. Third, there are inconveniences for students faced with using, and first learning to use, several different departmental systems. Of course, this difficulty can be overcome by requiring departments to purchase compatible systems and by interconnecting all systems.

Many academic computing specialists expect interconnected departmental computers to become the dominant means of academic computing in the next decade.

Personal computers

The recent development of personal microcomputers provides another alternative for instructional computing. Very low unit costs (one or two thousand dollars) make computing possible for departments otherwise unable to afford or gain authorization for large facilities. Microcomputer facilities suffer from many of the same problems as minicomputer facilities. In addition, microcomputers are limited in the services they provide, are slower than their large competitors, and may not be designed for rugged use by large groups of students. Still they can prove quite adequate for elementary applications. Further, by being less intimidating and more exciting than larger computers, they can play a role in overcoming a student's "computer anxiety."

Terminals

Terminals are used for remote, interactive access to large computers. Some have small memories and primitive editing capabilities. Departments often have a greater choice in selecting terminals to connect to computer systems than they do in selecting the systems themselves. Cost, speed, and durability are primary factors influencing the selection of a terminal. By these criteria, video terminals are preferable. The availability of graphical output and local editing features are other factors to consider when choosing terminals. Hard-copy (printing) terminals are more expensive and tend to be slower than video terminals, but they do provide users with a permanent record of their work, and so some printing terminals are necessary (medium or high speed printers can be used in conjunction with video terminals to provide this record). Video terminals may also be used in conjunction with television monitors to provide classroom displays of computer output. For such output to be visible in a large classroom, either many monitors must be provided or the video terminals employed must use larger, and hence fewer, characters in their display.

Applications

The suitability of a particular computing facility depends most upon its intended applications. The rest of this section discusses the most common academic uses of computers and how well different types of computing facilities serve these uses.

Introductory programming

Any of the three types of facilities can serve as a vehicle for teaching beginners to program and for introducing computational examples into elementary mathematics courses. Such uses typically involve large numbers of students writing relatively simple programs. Larger facilities tend to provide a greater choice of programming languages, although modern languages such as PASCAL and PL/I are becoming increasingly available even on microcomputers. Larger machines tend to be faster also; even though use of such machines is shared, students will find that they process simple programs much faster than microcomputers. Costs, however, tend to be roughly equal for simple interactive computing on the three types of facilities—around $2.00 per hour. These costs can be reduced significantly by using larger machines in a noninteractive, batch-processing mode. This mode of use, while predominant in the past, is becoming less popular as minicomputers and microcomputers make a more responsive computing environment available and affordable.

Advanced programming

Advanced programming is more distinguished from introductory programming in its requirements for more sophisticated languages and for facilities to handle large programs. Microcomputers at present do not meet these requirements; the languages they provide are quite restrictive, and large programs exceed their capacity. Execution times and costs for large programs tend to be lowest on large machines under batch processing, but minicomputers are becoming competitive both in price and speed.

Program development and maintenance

Program development is influenced heavily by the computing environment in which it occurs. Convenient interactive editing capabilities accelerate the task of writing and correcting a program; microcomputers, with almost instantaneous response, do a particularly good job of editing. Facilities for file storage enable program development to be spread over several sessions. Large machines provide less expensive storage and much faster retrieval of information; they also facilitate sharing programs among users and provide centralized backup. Microcomputer facilities can distribute the costs of file storage by requiring users to purchase individual floppy disks, but unless a centralized store is provided through a network, sharing information can be difficult.

Graphics

One of the primary attractions of personal microcomputers is their ability to generate graphic displays and to enable users to interact with these displays. Larger systems, unless specifically tailored to graphic applications, tend to have primitive graphic facilities at best.

APPLICATION PACKAGES

Application packages available for various machines provide aids for numerical and symbolic computations. Typical areas of application include statistics, linear programming, numerical solution of differential equations, and algebraic formula manipulation. Such packages are more widely available on larger machines. Large computations often require an unacceptably long time on microcomputers (several hours) and may exceed the memory size of small computers.

MISCELLANEOUS APPLICATIONS

Word processing systems facilitate production of course notes, research papers, and term papers. If good word processing facilities are available, they are likely to quickly generate heavy faculty use. Simple word processing software is available for personal computers, but a minicomputer (or powerful $5,000-plus microcomputer) is needed for good mathematically-oriented word processing software, such as the UNIX system. Large computers often have poor word processing capabilities.

Data base systems are of more use in the social sciences than in the mathematical sciences, but can be used to provide real data for analysis in statistics courses. Such systems require a centralized file store on a larger computer.

Real-time data acquisition is of interest in the natural sciences. They can also be used to provide real data for mathematical analysis. Dedicated microcomputers are better suited to laboratory instrumentation than are shared machines.

ACM Curriculum 78

The following computer science course syllabi are reproduced from the ACM Curriculum 78 Report in *Communications of ACM,* March 1979, pp. 147-166. (Copyright 1979, Association for Computing Machinery, Inc.) They provide eight core courses for a computer science major.

CS1. Computer Programming I

OBJECTIVES:

- To introduce problem solving methods and algorithm development;
- To teach a high-level programming language that is widely used; and

- To teach how to design, code, debug, and document programs using techniques of good programming style.

COURSE OUTLINE:

The material on a high-level programming language and on algorithm development can be taught best as an integrated whole. Thus the topics should not be covered sequentially. The emphasis of the course is on the techniques of algorithm development and programming with style. Neither esoteric features of a programming language nor other aspects of computers should be allowed to interfere with that purpose.

TOPICS:

A. *Computer Organization.* An overview identifying components and their functions, machine and assembly languages. (5%)
B. *Programming Language and Programming.* Representation of integers, real, characters, instructions. Data types, constants, variables. Arithmetic expression. Assignment statement. Logical expression. Sequencing, alternation, and iteration. Arrays. Subprograms and parameters. Simple I/O. Programming projects utilizing concepts and emphasizing good programming style. (45%)
C. *Algorithm Development.* Techniques of problem solving. Flowcharting. Stepwise refinement. Simple numerical examples. Algorithms for searching (e.g., linear, binary), sorting (e.g., exchange, insertion), merging of ordered lists. Examples taken from such areas as business applications involving data manipulation, and simulations involving games. (45%)
D. *Examinations.* (5%)

CS2. Computer Programming II

OBJECTIVES:

- To continue the development of discipline in program design, in style and expression, in debugging and testing, especially for larger programs;
- To introduce algorithmic analysis; and
- To introduce basic aspects of string processing, recursion, internal search/sort methods and simple data structures.

PREREQUISITE: CS 1.

COURSE OUTLINE:

The topics in this outline should be introduced as needed in the context of one or more projects involving larger programs. The instructor may choose to begin with the statement of a sizable project, then utilize structured programming techniques to develop a number of small projects each of which involves string processing, recursion, searching and sorting, or data structures. The emphasis on good programming style, expression, and documentation, begun in CS1, should be continued. In order to do this effectively, it may be necessary to introduce a second language (especially if a language like Fortran is used in CS1). In that case, details of the language should be included in the outline. Analysis of algorithms should be introduced, but at this level such analysis should be given by the instructor to the student.

Consideration should be given to the implementation of programming projects by organizing students into programming teams. This technique is essential in advanced level courses and should be attempted as early as possible in the curriculum. If large class size makes such an approach impractical, every effort should be made to have each student's programs read and critiqued by another student.

TOPICS:

A. *Review.* Principles of good programming style, expression, and documentation. Details of a second language if appropriate. (15%)
B. *Structured Programming Concepts.* Control flow. Invariant relation of a loop. Stepwise refinement of both statements and data structures, or top-down programming. (40%)
C. *Debugging and Testing.* (10%)
D. *String Processing.* Concatenation. Substrings. Matching. (5%)
E. *Internal Searching and Sorting.* Methods such as binary, radix, Shell, quicksort, merge sort. Hash coding. (10%)
F. *Data Structures.* Linear allocation (e.g., stacks, queues, deques) and linked allocation (e.g., simple linked lists). (10%)
G. *Recursion.* (5%)
H. *Examinations.* (5%)

CS3. Introduction to Computer Systems

OBJECTIVES:

- To provide basic concepts of computer systems;
- To introduce computer architecture; and
- To teach an assembly language.

PREREQUISITE: CS 2.

COURSE OUTLINE:

The extent to which each topic is discussed and the ordering of topics depends on the facilities available and the nature and orientation of CS4 described below. Enough assembly language details should be covered and projects assigned so that the student gains experience in programming a specific computer. However, concepts and techniques that apply to a broad range of computers should be emphasized. Programming methods that are developed in CS1 and CS2 should also be utilized in this course.

TOPICS:

A. *Computer Structure and Machine Language.* Memory, control, processing and I/O units. Registers, principal machine instruction types and their formats. Character representation. Program control. Fetch-execute cycle. Timing. I/O Operations. (15%)
B. *Assembly Language.* Mnemonic operations. Symbolic addresses. Assembler concepts and instruction format. Data-word definition. Literals. Location counter. Error flags and messages. Implementation of high-level language constructs. (30%)
C. *Addressing Techniques.* Indexing. Indirect Addressing. Absolute and relative addressing. (5%)
D. *Macros.* Definition. Call. Parameters. Expansion. Nesting. Conditional assembly. (10%)
E. *File I/O.* Basic physical characteristics of I/O and auxiliary storage devices. File control system. I/O specification statements and device handlers. Data handling, including buffering and blocking. (5%)
F. *Program Segmentation and Linkage.* Subroutines. Coroutines. Recursive and re-entrant routines. (20%)
G. *Assembler Construction.* One-pass and two-pass assemblers. Relocation. Relocatable loaders. (5%)
H. *Interpretive Routines.* Simulators. Trace. (5%)
I. *Examinations.* (5%)

CS4. Introduction to Computer Organization

OBJECTIVES:

- To introduce the organization and structuring of the major hardware components of computers;
- To understand the mechanics of information transfer and control within a digital computer system; and
- To provide the fundamentals of logic design.

PREREQUISITE: CS 2.

COURSE OUTLINE:

The three main categories in the outline, namely computer architecture, arithmetic, and basic logic design, should be interwoven throughout the course rather than taught sequentially. The first two of these areas may be covered, at least in part, in CS3 and the amount of material included in this course will depend on how the topics are divided between the two courses. The logic design part of the outline is specific and essential to this course. The functional, logic design level is emphasized rather than circuit details which are more appropriate in engineering curricula. The functional level provides the student with an understanding of the mechanics of information transfer and control within the computer system. Although much of the course material can and should be presented in a form that is independent of any particular technology, it is recommended that an actual simple minicomputer or microcomputer system be studied. A supplemental laboratory is appropriate for that purpose.

TOPICS:

A. *Basic Logic Design.* Representation of both data and control information by digital (binary) signals. Logic properties of elemental devices for processing (gates) and storing (flipflops) information. Description by truth tables, Boolean functions and timing diagrams. Analysis and synthesis of combinatorial networks of commonly used gate types. Parallel and serial registers. Analysis and synthesis of simple synchronous control mechanisms; data and address buses; addressing and accessing methods; memory segmentation. Practical methods of timing pulse generation. (25%)

B. *Coding.* Commonly used codes (e.g., BCD, ASCII). Parity generation and detection. Encoders, decoders, code converters. (5%)

C. *Number Representation and Arithmetic.* Binary number representation, unsigned addition and subtraction. One's and two's complement, signed magnitude and excess radix number representations and their pros and cons for implementing elementary arithmetic for BCD and excess-3 representations. (10%)

D. *Computer Architecture.* Functions of, and communication between, large-scale components of a computer system. Hardware implementation and sequencing of instruction fetch, address construction, and instruction execution. Data flow and control block diagrams of a simple processor. Concept of microprogram and analogy with software. Properties of simple I/O devices and their controllers, synchronous control, interrupts. Modes of communications with processors. (35%)

E. *Example.* Study of an actual, simple minicomputer or microcomputer system. (20%)

F. *Examinations.* (5%)

CS5. Introduction to File Processing

OBJECTIVES:

- To introduce concepts and techniques of structuring data on bulk storage devices;
- To provide experience in the use of bulk storage devices; and
- To provide the foundation for applications of data structures and file processing techniques.

PREREQUISITE: CS 2.

COURSE OUTLINE:

The emphasis given to topics in this outline will vary depending on the computer facilities available to students. Programming projects should be assigned to give students experience in file processing. Characteristics and utilization of a variety of storage devices should be covered even though some of the devices are not part of the computer system that is used. Algorithmic analysis and programming techniques developed in CS2 should be utilized.

TOPICS:

A. *File Processing Environment.* Definitions of record, file, blocking, compaction, database. Overview of database management system. (5%)

B. *Sequential Access.* Physical characteristics of sequential media (tape, cards, etc.). External sort/merge algorithms. File manipulation techniques for updating, deleting and inserting records in sequential files. (30%)

C. *Data Structures.* Algorithms for manipulating linked lists. Binary, *B*-trees, *B**-trees, and *AVL* trees. Algorithms for transversing and balancing trees. Basic concepts of networks (plex structures). (20%)

D. *Random Access.* Physical characteristics of disk, drum, and other bulk storage devices. Algorithms and techniques for implementing inverted lists, multilist, indexed sequential, and hierarchical structures. (35%)

E. *File I/O.* File control systems and utility routines, I/O specification statements for allocating space and cataloging files. (5%)

F. *Examinations.* (5%)

CS6. Operating Systems & Comp. Architecture

OBJECTIVES:

- To develop an understanding of the organization and architecture of computer systems at the register-transfer and programming levels of system description;

- To introduce the major concept areas of operating systems principles;
- To teach the inter-relationships between the operating system and the architecture of computer systems.

PREREQUISITES: CS3 AND CS4.

COURSE OUTLINE:

This course should emphasize concepts rather than case studies. Subtleties do exist, however, in operating systems that do not readily follow from concepts alone. It is recommended that a laboratory requiring hands-on experience be included with this course.

The laboratory for the course would ideally use a small computer where students could actually implement sections of operating systems and have them fail without serious consequences to other users. This system should have, at a minimum, a CPU, memory, disk or tape, and some terminal device such as a teletype of CRT. The second best choice for the laboratory experience would be a simulated system running on a larger machine.

The course material should be liberally sprinkled with examples of operating system segments implemented on particular computer system architectures. The interdependence of operating systems and architecture should be clearly delineated. Integrating these subjects at an early stage in the curriculum is particularly important because the effects of computer architecture on systems software has long been recognized. Also, modern systems combine the design of operating systems and the architecture.

TOPICS:

A. *Review.* Instruction sets. I/O and interrupt structure. Addressing schemes. Microprogramming. (10%)
B. *Dynamic Procedure Activation.* Procedure activation and deactivation on a stack, including dynamic storage allocation, passing value and reference parameters, establishing new local environments, addressing mechanics for accessing parameters (e.g., displays, relative addressing in the stack). Implementing non-local references. Re-entrant programs. Implementation on register machines. (15%)
C. *System Structure.* Design methodologies such as level, abstract data types, monitors, kernels, nuclei, networks of operating system modules. Proving correctness. (10%)

D. *Evaluation.* Elementary queueing, network models of systems, bottlenecks, program behavior, and statistical analysis. (15%)
E. *Memory Management.* Characteristics of the hierarchy of storage media, virtual memory, paging, segmentation. Policies and mechanisms for efficiency of mapping operations and storage utilization. Memory protection. Multiprogramming. Problems of auxiliary memory. (20%)
F. *Process Management.* Asynchronous processes. Using interrupt hardware to trigger software procedure calls. Process stateword and automatic SWITCH instructions. Semaphores. Ready lists. Implementing a simple scheduler. Examples of process control problems such as deadlock, product/consumers, readers/writers. (20%)
G. *Recovery Procedures.* Techniques of automatic and manual recovery in the event of system failures. (5%)
H. *Examinations.* (5%)

CS7. Data Structures and Algorithm Analysis

OBJECTIVES:

- To apply analysis and design techniques to non-numeric algorithms which act on data structures;
- To utilize algorithmic analysis and design criteria in the selection of methods for data manipulation in the environment of a database management system.

PREREQUSITES: CS5.

COURSE OUTLINE:

The material in this outline could be covered sequentially in a course. It is designed to build on the foundation established in the elementary material, particularly on that material which involves algorithm development and data structures and file processing. The practical approach in the earlier material should be made more rigorous in this course through the use of techniques for the analysis and design of efficient algorithms. The results of this more formal study should then be incorporated into data management system design decisions. This involves differentiating between theoretical or experimental results for individual methods and the results which might actually be achieved in systems which integrate a variety of methods and data structures. Thus, database management systems provide the applications environment for topics discussed in the course.

Projects and assignments should involve implementation of theoretical results. This suggests an alternative way of covering the material in the course; namely, to treat concepts, algorithms, and analysis in class and deal with their impact on system design in assignments. Of course, some in-class discussions of this impact would occur, but at various times throughout the course rather than concentrated at the end.

TOPICS:

A. *Review.* Basic data structures such as stacks, queues, lists, trees. Algorithms for their implementation. (10%)

B. *Graphs.* Definition, terminology, and property (e.g., connectivity). Algorithms for finding paths and spanning trees. (15%)

C. *Algorithms Design and Analysis.* Basic techniques of design and analysis of efficient algorithms for internal and external sorting/merging/searching. Intuitive notions of complexity (e.g., NP-hard problems). (30%)

D. *Memory Management.* Hashing. Algorithms for dynamic storage allocation (e.g., buddy system, boundary-tag), garbage collection and compaction. (15%)

E. *System Design.* Integration of data structures, sort/merge/search methods (internal and external) and memory media into a simple database management system. Accessing methods. Effects on run time, costs, efficiency. (25%)

F. *Examinations.* (5%)

CS8. Organization of Programming Languages

OBJECTIVES:

- To develop an understanding of the organization of programming languages, especially the run-time behavior of programs;
- To introduce the formal study of programming language specification and analysis;
- To continue the development of problem solution and programming skills introduced in the elementary level material.

PREREQUISITES: CS2; RECOMMENDED: CS3, CS5.

COURSE OUTLINE:

This is an applied course in programming language constructs emphasizing the run-time behavior of programs. It should provide appropriate background for advanced level courses involving formal and theoretical aspects of programming languages and/or the compilation process.

The material in this outline is not intended to be covered sequentially. Instead, programming languages could be specified and analyzed one at a time in terms of their features and limitations based on their run-time environments. Alternatively, desirable specification of programming languages could be discussed and then exemplified by citing their implementations in various languages. In either case, programming exercises in each language should be assigned to emphasize the implementations of language features.

TOPICS:

A. *Language Definition Structure.* Formal language concepts including syntax and basic characteristics of grammars, especially finite state, context-free, and ambiguous. Backus-Naur Form. A language such as Algol as an example. (15%)

B. *Data Types and Structures.* Review of basic data types, including lists and trees. Constructs for specifying and manipulating data types. Language features affecting static and dynamic data storage management. (10%)

C. *Control Structures and Data Flow.* Programming language constructs for specifying program control and data transfer, including DO ...FOR, DO ...WHILE, REPEAT ...UNTIL, BREAK, subroutines, procedures, block structures, and interrupts. Decision tables, recursion. Relationship with good programming style should be emphasized. (15%)

D. *Run-time Consideration.* The effects of run-time environment and binding time on various features of programming languages. (25%)

E. *Interpretative Languages.* Compilation vs. interpretation. String processing with language features such as those available in SNOBOL 4. Vector processing with language features such as those available in SPL. (20%)

F. *Lexical Analysis and Parsing.* An introduction to lexical analysis including scanning, finite state acceptors and symbol tables. An introduction to parsing and compilers including push-down acceptors, top-down and bottom-up parsing. (10%)

G. *Examinations.* (5%)

Subpanel Members

ALAN TUCKER, CHAIR, SUNY-Stony Brook.
GERALD ENGEL, Christopher Newport College.
STEPHEN GARLAND, Dartmouth College.
BERT MENDELSON, Smith College.
ANTHONY RALSTON, SUNY-Buffalo.

Modeling and Operations Research

This chapter contains the report of the Subpanel on Modeling and Operations Research of the CUPM Panel on a General Mathematical Sciences Program, reprinted with minor changes from Chapter V of the 1981 CUPM report entitled RECOMMENDATIONS FOR A GENERAL MATHEMATICAL SCIENCES PROGRAM.

Experience in Applications

This chapter is concerned with mathematical modeling and associated interactive and experience-oriented approaches to teaching mathematical sciences. Mathematical modeling attempts to involve students in the more creative and early design aspects of problem formulation, as well as provide them with a more complete exposure to how mathematics interfaces with other activities in solving problems arising outside of mathematics itself. Model building is a major ingredient of operations research and the contemporary uses of mathematics in the social, life and decision sciences. In addition to being important in their own right, these newer uses of mathematics provide a rich source of suitable materials for interaction and modeling which complement the many modern and classical applications of mathematics in the physical sciences and engineering.

This chapter is intended to assist mathematics faculty in implementing the main panel's recommendation that mathematical sciences majors should have substantial experience with mathematical modeling. Subsequent sections discuss the modeling process in some detail; provide specific suggestions for conducting student projects, applications-experience-related courses and other such programs, along with general recommendations concerning modeling courses at different levels; explain the field of operations research and the requirements for graduate study. The final two sections present outlines for four courses in operations research and modeling, and a compendium of resources and references for modeling courses.

Learning and doing mathematics is a rather individualized and personal activity. The typical classroom lecture in which students are passive spectators has obvious limitations. Students need supervised hands-on experience in problem solving and constructing rigorous proofs. A large variety of alternate teaching techniques and special programs have been developed in attempts to meet this need. These include problem solving approaches using materials from pure and applied mathematics, such as the methods of G. Pólya and R.L. Moore. Problem solving teams for competitions such as the Putnam contest and special departmental practica exist in many colleges. Special courses or seminars on modeling, case studies, and project-oriented activity are becoming more common, as are mathematics clinics and consulting bureaus. Co-op and work-study programs, summer internships, and various other student exchanges have been successfully implemented at some institutions.

The Modeling Subpanel believes that applications and modeling should be included in a nontrivial way in most college-level mathematical sciences courses. Concern with applications has been an important historical force and a major cultural ingredient in the development of all mathematics. Further, the Modeling Subpanel strongly recommends that all mathematical sciences students should obtain first-hand experience with realistic applications of mathematics from the initial stage of model formulation through interpretation of solutions. This can be done in a project-oriented modeling course in one of the alternate out-of-class modes mentioned above. Such an experience yields insight into the place of mathematics in the larger realm of science. It provides an appreciation for the need for interdisciplinary interaction and the limits of specialization. It offers a chance for individuals to make use of their own intuition and creative abilities, to sense the great joy of personal accomplishment, and to develop the confidence to confront similar problems after graduation. Finally, such experience may assist students in choosing careers and fields for future study.

Mathematical Modeling

Modeling is a fundamental part of the general scientific method and is of primary importance in applied mathematics. A model is a simpler realization or an idealization of some more complex reality created for the purpose of gaining new knowledge about a real situation by investigating properties and implications of the model. Models may take many different forms, from physical miniatures to pure intellectual substitutes. Study of a model will hopefully provide understanding and new information about real phenomena which are too complex, excessively expensive, or impossible to analyze in their original setting.

We tend to take the amazing effectiveness of models for granted today. The reader should give a moment's thought to the following examples. One can learn a great deal about a proposed aircraft from wind tunnel experiments before building a costly prototype, and one can learn much about flying an existing airplane from a computer-aided cockpit simulator. Simple computer simulations can provide insights into the complex flow or queueing behavior of traffic in a transportation system. Theoretical studies about elementary particles have provided new insight into fundamental physical laws and have guided subatomic experimentation.

right side from "mathematical model" to "mathematical solution." This is the deductive activity of finding solutions to well-formulated mathematical problems. It is usually the most logical, well-defined and straightforward part of modeling, although not necessarily the easiest. It is often the most immediately pleasing, elegant, and intellectual part. This "side" of the "modeling square" is the one covered best in standard applied mathematics courses. Unfortunately, most teaching of applied mathematics is confined to discussing just model-solving mathematical techniques, with superficial treatment of the other three sides of the square, whereas these other sides often involve much more creativity, interaction with other disciplines, and communication skills.

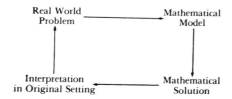

A Simple Model of Mathematical Modeling

Figure 1

The process of mathematical modeling can be simply represented with the diagram in Figure 1. One begins with a problem which arises more or less directly out of the "real world." One builds an abstract model for purposes of analysis, and this frequently takes a mathematical form. The model is solved in this abstract setting. The solution is then interpreted back into its original context. Finally, the analytical conclusions are compared with reality. If they fall short of matching the real situation, then modifications of the model may be called for, and one proceeds around this cycle again. One often proceeds back and forth within a cycle and makes successive iterations about this figure many times before arriving at a satisfactory representation of the real world.

The creation of new knowledge via this modeling route is at the heart of theoretical science and applied mathematics. We will use the word "modeling" to describe the complete progress illustrated in Figure 1. Frequently this term is used only for the model formulation step (the top arrow in the figure). A full discussion of the four steps in this modeling paradigm follow. Additional steps refining the modeling process are sometimes inserted; for example, see Figure 2.

First consider the downward pointing arrow on the

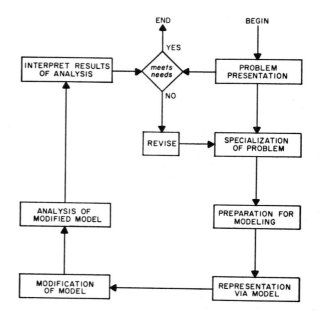

A Refined Model of Mathematical Modeling

Figure 2

The bottom arrow in Figure 1 is concerned with translating or explaining a purely mathematical result in terms of the original real world setting. This involves the need to communicate in a precise and lucid manner. (Inexperience in this skill, according to many employers, is a serious shortcoming in mathematics graduates). This aspect of a mathematical scientist's training should not be left to courses in other sciences or to on-the-job learning after graduation.

In describing the meaning of a mathematical solution one must take great care to be complete and honest. It is dangerous to discard quickly some mathematical solutions to a physical problem as extraneous or having no physical meaning; there have been too many historical incidences where "extraneous" solutions were of fundamental importance. Likewise, one should not select out just the one preconceived answer which the "boss" is looking for to support his or her position. A decision maker frequently does not want just one optimal solution, but desires to know a variety of "good" solutions and the range of reasonable options available from which to select.

There is an old adage to the effect that bosses do not act on quantitative recommendations unless they are communicated in a manner which makes them understandable to such decision makers. This communication can often be a difficult task because of the technical nature of the formulation and solution, and also because large quantities of data and extensive computation may need to be compressed to a manageable size for the layman to understand in a relatively limited time. If mathematical education gave more attention to this aspect of mathematical modeling, there might be wider recognition and visibility of mathematicians in society beyond the academic world!

A major step in real world modeling is to validate models critically and to check out solutions against the original phenomena and known results. This step, represented by the left upward arrow in Figure 1, may involve experimentation, verifying, and evaluating. Two major criteria for evaluating a model are simplicity and accuracy of prediction. Questions about the range of validity, sensitivity of parameters errors resulting from approximations, and such should be investigated. In many cases, a modeling project will simply confirm from another perspective properties that are already believed to be true. The real gain from modeling activity occurs when the modeling leads to discovery of new knowledge (which subsequently is confirmed by other methods).

Modern mathematics education rarely involves itself with this left hand side of the modeling process, except perhaps for an occasional "eyeballing" of an answer or in projects undertaken by a mathematics or statistics consulting clinic. By omitting this activity, mathematical education misses an opportunity to become involved with real-world decision making, judgmental inputs, the limitations of its mathematical tools, and other more human aspects of science, as well as the reward of witnessing the acceptance of a new theory.

Finally, consider the top arrow in Figure 1 which represents the heart of the modeling activity. The construction of an abstract model from a real situation is the really creative activity and an important component of all theoretical science. Building models involves translating into mathematics, maintaining the essential ingredients while filtering out a great amount of excess baggage, and arriving at realistic and manageable intellectual limitations. The three basic elements of a model are:

1. A logical mathematical structure such as calculus, probability, or game theory;
2. An appropriate interpretation of the variables in that structure in terms of the given problem; and
3. A characterization with the structure of all laws and constraints pertinent to the problem.

To build such a mental construct, one must conceptualize, idealize and identify properties precisely. A model builder must carefully balance the tradeoffs between coarse simplifications and unnecessary details—often the effects of such tradeoffs are not apparent until subsequent validation (three steps later in the modeling process).

This initial part of modeling is clearly the most essential and valuable part of the whole process. It is usually the most difficult part. Eddington said "I regard the introductory part of a theory as the most difficult, because we have to use our brains all of the time. Afterwards we can use mathematics." Model building is an art, and must be taught as such.

An Undergraduate Modeling Course

This section discusses various approaches to designing mathematical sciences courses concentrating on the modeling process. The resources listed at the end of this chapter contain a wealth of additional information on models, the modeling process and specific modeling courses as well as references to supplementary materials which the reader may find useful in course design.

Practitioners in the physical or social sciences or engineering have an instinctive feeling of what the modeling process is all about, even if they are not able to articulate it well. Modeling is an important part of their work-a-day activity. For the most part, however, they prefer to leave the analysis and structure of the modeling process itself to workers in other disciplines, like mathematics, or to philosophers of science who are trying to understand the abstract theories underlying these results and how scientists get their results.

How does one go about acquiring experience in real-world modeling? The wrong place to start is looking at big models in the scientific literature which are broad in scope and the epitome of their kind. Indeed, one could probably learn more about sculpting by looking at the pieces that Michelangelo discarded than by looking at the Pieta. The mathematical techniques with which one is familiar will be a primary limiting factor in understanding models. Another factor is that real-world problem areas have their own peculiar "empirical laws" and "principles" which are commonly known to specialists in an area but are not easily accessible to the casual reader.

Apprentice modelers need some help and guidance in selecting model areas for study which will build their modeling skill without discouraging progress. The ideal way to do this within the college curriculum is to begin the modeling process as early as possible in the student's career and reinforce modeling over the entire period of study. That is, the modeling process should be an integral part of the curriculum. Most mathematics departments, for a variety of reasons, are not prepared to give modeling such a major emphasis. For them, a more reasonable approach is to design a course specifically around the modeling process.

Efforts to emphasize the modeling process in undergraduate courses on a broad scale began in the 1960's and were promoted mainly by engineers, operations researchers and social scientists. Extensive discussions of modeling in mathematics courses developed later. The modeling process has been brought into the classroom in many ways but two particular approaches are worth describing in some detail.

First there is the case study approach in which the modeling process is described in a series of examples that are more-or-less self-contained. The examples selected by the instructor are designed to bring out the basic features of the modeling process as well as to inform the students about basic models within a discipline. An excellent early example is *You and Technology: A High School Case Study Text* developed by the engineering departments of the PCM Colleges (Chester, PA), edited by N. Damaskos and M. Smyth.

The second approach applies "hands-on" experience to problems that may only be vaguely described. This approach is sometimes called "open-ended" or "experiential," because it is not clear at the outset what kind of a model will be successful in analyzing a problem, or indeed whether a particular problem is well-posed in any sense. An interesting sidelight on this approach to teaching the modeling process is that the models proposed by students for a particular problem depend not only on the students' breadth of knowledge but, as much as anything else, on time constraints and computer (and other) resources available. Engineers popularized the experiential approach in the early sixties with the high school program *Man Made World*, mostly as a means of exposing students at an early stage to engineering as a profession (a text of the same name was written for this program by J. Truxal, et al., McGraw-Hill publisher).

A range of courses emphasizing the modeling process is clearly possible between the case study approach and the experiential approach.

It is important to note that the scope of the engineering approach to modeling is much broader than just the technical aspects of the problem at hand. In designing a solution to a problem, engineers must take into account time constraints and build into their models prescribed economic and other technical constraints as well as consideration of the impact of their design on society. Engineers do not build elaborate models to explain the fundamental workings of nature nor do they seek the best possible solution to a problem in the absence of the proposed application of that solution. In spite of these differences, there is obviously a large overlap between the engineering and mathematical approaches to modeling.

We now characterize the components of a modeling course in a way that readers should find useful in designing a course to fit their own local needs. The Table on pp. 46-47 organizes much of this information for easy reference. There are six basic aspects of teaching modeling that must be considered:

1. Prerequisites. For whom is the course intended?
2. Effort level. How long—a few weeks, a semester, a year?
3. Course format. Experiential or case study approach? Team or individual work? Instructor's role. Communication skills used.
4. Resources available. Computer system, remote access, good software packages (students should become familiar with using some major software package). Access to expertise in fields considered. Appropriate handouts to keep students progressing.
5. Source of problems. Real-world or contrived? Open-ended or can student answer all questions by looking them up in the literature?
6. Technical thrust. What technical areas should the course emphasize, or avoid? Continuous or discrete models? Deterministic or stochastic? Role of computer programming.

We now expand a little on two of these components, effort level and course format. The level of effort devoted to a modeling course can range from "mini-projects," using a team approach to short projects within an established course, to major projects which last an entire year. The mini-project format requires a great deal of organization and preparation to make it work. See Borrelli and Busenburg "Undergraduate Classroom Experiences in Applied Mathematics" (*UMAP Journal*, Volume 1, 1980) for one approach to structuring a mini-project program, together with its pro's and con's. The one-semester case study course, judging from its popularity, is the best understood and trusted of modeling courses. There are good textbooks and a great many modules written for use in such a course (see list at end of chapter).

While most case studies texts on mathematical modeling are designed for upper-level courses, the text *You and Technology* (mentioned above), supplemented with modules, can easily be adapted for use in a freshman case studies course. Such a course might also present an opportunity for students to see the fundamental differences between engineering and mathematical approaches to modeling (this issue is treated nicely in *You and Technology*). An extensive outline is provided below for a special custom-made, lower-level modeling course.

Experiential modeling courses are not used as often as case study courses. Since the experiential approach is typically used on open-ended problems where the outcome is difficult to predict in advance, this approach is especially risky for a mathematics instructor who is teaching a modeling course for the first time. Nevertheless, experiences of various colleges over the last several years show that the experiential approach is feasible and that, whatever happens, students and instructors find it a rewarding experience. Several successful formats for experiential modeling courses have emerged. All seem to use the team approach with occasional guidance by consultants, as needed. It should be noted that many industrial employers treat such experiential modeling as job-related experience in assessing a student's job qualifications. References at the end of this chapter contain descriptions of the well-known Mathematics Clinics in Claremont and other experiential modeling courses (interested readers can write directly to Harvey Mudd College for first-hand advice).

We close this Section with some important general points to keep in mind when designing any modeling course.

- To encourage initiative and independent work, students should have access to, and be responsible for using, support resources such as documentation of software and previous student projects.
- If high standards are imposed on writing of reports, then these reports deserve some exposure; they should not just be shoved in filing cabinets and forgotten. Instructors should encourage students to seek publication of a paper based on their reports, if warranted, or an article in the campus newspaper. Abstracts of recent reports should be made available to students early in a modeling course. When students know their work will get exposure, they are motivated to write good reports.
- It is valuable to integrate the modeling process into the curriculum as widely as possible and not just as an add-on special course with no connection to any other mathematical sciences course.
- A problem with most modeling courses is that the material in them quickly becomes dated. When students discover that they are working on the same projects or models as their classmates did last year, they lose enthusiasm. What is needed is a format for automatically updating the material. A constant flow of real-world problems, as come into a mathematics consulting clinic, is a great advantage.

Operations Research

Operations research is a mathematical science closely connected to mathematical modeling. Although some notable contributions were made prior to 1940, operations research grew out of World War II. The analysis of military logistics, supply and operational problems by scientists from many different disciplines generated the techniques and approaches that evolved into modern operations research. This subject studies complex systems, structures and institutions with a view towards operating such multiparameter systems more efficiently within various constraints, such as scarce resources. Operations research analyses are used to optimize current activities and predict future feasibility. The complexity of its problems has made operations research heavily dependent on high-speed digital computers. It is now used in fields in which decisions were traditionally made on the basis of less quantitative approaches, such as "experience" or mere hunches. There is frequently a major concern with "people" as well as "things," and the man-system interface in a complex social activity. Major national concerns such as productivity, environmental impact and energy supply have a large operations research component.

The approach in operations research is multidisciplinary in nature, and uses common sense, data, and substantial empiricism (heuristics) combined with new, as well as repackaged traditional, mathematical methodologies. The principal mathematical theories of operations research are mathematical programming and stochastic processes. Major topics in these theories are mentioned in the operations research course contents in the next section. Operations research has major overlap with the fields of industrial engineering, management science, mathematical economics, econometrics and decision theory.

Anatomy of a Modeling Course

Ingredients	Background and Source Material	Remarks
PREREQUISITES:		
Lower Division. Single variable calculus, a science course with lab, some computing.	Case study approach most likely. See, e.g., "You and Technology" or suitable UMAP modules.	If the team approach is selected then there can be some flexibility in these prerequisites.
Upper Division. Multivariable calculus, linear algebra, computation and some computer programming, basic prob/stat., some diff. eqns., a science course with lab.	For experiential approach and case study approach consult appropriately noted reference.	If modeling course is not required, then some thought must be given as to how students can be attracted to such a course: descriptions in registration packets, posters, note to advisor, etc.
EFFORT LEVEL:		
Partial Course. Recommended minimum of 2 weeks out of a 3 hour course preceded by a tooling up period.	Mini-projects are a possibility here. See Borelli and Busenberg. Format of mini-projects can be effectively structured. See Becker, *et al.*, "Handbook for Projects."	Important that mini-project work *not* be simply added to standard load of the host course—it should replace some required work; e.g., an exam.
Full Course. May be designed to fit into special options, either to give job-related training or introduction to modeling process with important models in a discipline.	Many possibilities exist for modeling courses for a full semester—see items below. For a discussion of pros and cons, see Borelli and Busenberg.	Format of instruction can seriously affect the student's interest as well as his capacity for effective work—see "Format" section below for possibilities.
COURSE FORMAT:		
Case Study. The modeling process presented via examples that are more-or-less self-contained.	Material selected from modules, textbooks, conference proceedings, or journals.	Advanced students can be asked to lecture on material that is well enough organized.
Experiential. Hands-on approach to open-ended projects incorporating the modeling process. Some possibilities are:		Internships, work-study programs not appropriate for inclusion here.
1. Problem-centered Course. Class divided into teams to work on a sequence of projects and share experience.	Needs highly experienced instructor to select and present the projects and watch over progress of the teams. Class size limited by instructor's energy. See Borelli and Busenberg for more details.	Oral presentation and written reports are emphasized. Most demanding of instructor's time.
2. Mathematics Clinic, Consulting Group. Intensive, industry-supported team effort on a single project, usually for one year.	Composition of team is critical. See Claremont Clinic Articles for details. Because of time constraints, able support staff must be readily available.	Team communication skills highly emphasized in Clinic program and is crucial to success. Team has main responsibility for work, instructor advises. Student handbook at Claremont Clinic (by Handa) available on request.
3. Research Assistance. Students aid faculty in research work.	MIT has a highly organized program which does this. Mostly, however, it's catch-as-catch-can. The Institute of Decision Science, Claremont Men's College, has developed a classroom approach to such work.	A danger here is that the success of the faculty member's research may take precedence over the impact on the students' education. Students' needs could get lost in the shuffle.
4. Mini-projects. Team approach on short projects within an established course.	See Borelli and Busenberg.	Emphasizes writing skills, highly structured activity; see "Handbook for Projects" by Becker, *et al.*

Anatomy of a Modeling Course

Ingredients	Background and Source Material	Remarks
Resources Available:		
Computer. Good access to a high level computer (preferably with time-sharing capability) having good software packages is very important for the success of most modeling courses.		Computer graphics capabilities and knowledgeable (and accessible) consultants at the computer center add not only a professional touch but also help teams live within their time constraints.
Experienced Consultants. Access to knowledgeable colleagues, experts in local industrial firms, and talented computer center personnel are all helpful in keeping a team's progress from faltering.	A successful, long-term program depends to a large extent on the Director's ability to secure *willing* assistance from able consultants.	Be sure that consultants help is acknowledged by the students in all written reports, even if it is only of a casual nature.
Supplemental Materials. Handouts on how to work in a team on projects, or where to go for help, etc., lessen the student's feeling of abandonment when working on projects.	For project work, see the Handbooks by Becker, *et al.,* Handa, Seven and Zagar, and for computer graphics, Saunders, *et al.* (all were developed at Harvey Mudd College and are available on request).	
Source of Problems:		
Real World. Open-ended problems submitted by local industrial firms or government agencies which are of current interest to them, or problems from current research of colleagues.	See Borelli and Spanier for a description of one effective method of recruiting sponsored projects from industry. MIT has a highly organized way of advertising current research of its faculty and laboratories and whether undergraduates can play a role or not.	Used only in experiential type modeling courses.
Contrived. Open-ended problems pulled from a variety of sources: from technical journals, suggestions from colleagues, books, etc.	The modeling books in the references are good sources of problems.	Used mostly in experiential type modeling course.
Case Studies. Reasonably well self-contained descriptions of completed projects or problems.	Good sources in modules, proceedings of conferences on case studies and books.	Used only for case study type of modeling course.
Technical Thrust:		
Discrete-OR. Problems whose models involve discrete structures, programming, or optimization within discrete settings. Also interpolation with finite structures in continuous settings.		Deterministic and stochastic methods are both possibilities here.
Continuous. Problems whose models involve differential or integral equations, continuous probabilities, or optimization within continuous setting.		
Computer. Problems with main goal the production of software either at the systems level or solvers for a class of equations in continuous settings, along with error analysis of same. For DEC users, the IMSL package is a good all-around one to have available on the system.		

There are many opportunities for mathematical sciences majors to pursue graduate studies or find employment in operations research and related fields. Industrial mathematicians in all fields find themselves faced with operations research problems from time-to-time. Thus it is important for mathematical sciences students to have some exposure to operations research and its applications, and also knowledge of its career possibilities. This classroom exposure to operations research can occur in conjunction with undergraduate modeling experience or in a specific course on operations research. The current relevance and naturalness of this subject are immediately clear to students, and realistic projects at various levels of difficulty are readily available. An interesting article by D. Wagner about operations research appeared in the *American Mathematical Monthly* (82, p. 895). Students should also be referred to the booklet *Careers in Operations Research,* available from the Operations Research Society of America, 428 Preston Street, Baltimore, MD 21202.

A student interested in graduate work in operations research should have a solid preparation in undergraduate core mathematics: calculus, linear algebra, real analysis, plus courses in probability, introductory computer science and modeling. A course in operations research itself is more important as a way to learn if one likes the field than as a prerequisite for graduate study. A substantial minor in a relevant area outside mathematics (as recommended for all mathematical sciences majors in the first chapter, "Mathematical Sciences") is important. This outside work should include a sampling of quantitative courses in the social sciences, business, or engineering (if available). Experience solving some problems involving substantial computer computation and an exposure to nontrivial algorithms are also desirable.

At some institutions, mathematics departments are now preparing to offer an operations research course for the first time, while other institutions may have many operations research courses offered in mathematics, economics, business, industrial engineering and computer science. In either extreme and situations in between, mathematical sciences students are best served by some form of interdepartmental cooperation, or at least coordination of offerings. If a mathematics department is planning to offer an operations research course when none previously existed at the institution, mathematics should work closely with other interested departments.

In planning this first course, mathematicians could seek contacts with local industry to obtain practitioners as visiting lecturers. On the other hand, an introductory operations research course can be taught by most college mathematics professors with appropriate attitudes if they are willing to undertake some self study. Indeed, faculty without formal operations research training who are going to teach such a course should be strongly encouraged to learn about the field by attendance at short courses, participation in a department seminar on the subject, or by sabbatical leave (or other released time) at universities or industrial laboratories with operations research activities.

Course Descriptions

Four sample courses on operations research and modeling are described below. Only more general remarks are given for the courses in operations research and stochastic processes since these have become fairly standardized in recent years. More specific details are provided for an elementary-level modeling course using discrete mathematics and for a more advanced modeling course using continuous methods. These are merely illustrations of the wide variety of different sorts of modeling courses which can be taught. The 1972 CUPM *Recommendations on Applied Mathematics* contain a detailed description of a physical-sciences oriented modeling course. Such a modeling course continues to be very valuable and in no way should be considered dated. Many basic intermediate-level courses in the physical sciences are also excellent modeling courses, from the point of view of a mathematical sciences major.

Introductory Operations Research

Much of the material in an introductory operations research course for undergraduates has become fairly standard. The course covers primarily deterministic methods. Most publishing companies have good introductory operations research texts (the text title may be Linear Programming, the course's main topic). The level of this course can vary depending on the prerequisites and student maturity. It is normally an upper-level offering with a prerequisite or corequisite of linear algebra. Calculus and probability should be required if stochastic models are also included.

An operations research course can be a "pure mathematics" course which stresses the fundamental properties of systems of linear inequalities, basic geometry of polyhedra and cones, discrete optimization and complexity of algorithms. Most operations research courses, however, emphasize the many applications which can be solved by linear programming and related techniques of combinatorial optimization. Such courses usually devote some time to efficient algorithms and practical numerical methods (to avoid roundoff errors), as well as basic notions of computational complexity. While problem solving and modeling are important, a first operations research course should cover some topic in reasonable depth and not be merely a collection of simple techniques and routine applications.

COURSE CONTENT

The course should start with a brief discussion of the general nature, history and philosophy of operations research. Some of the older texts such as *Introduction to Operations Research* by C. Churchman, R. Ackoff and E. Arnoff, Wiley, 1957, and *Methods of Operations Research* by P. Morse and G. Kimball, Wiley, 1951, devote extensive space to history. The instructor should not spend much time on history at the beginning of a course but instead should weave it into discussions throughout a course.

The first half of the course in usually devoted to linear programming: its theory, the simplex algorithm, and applications. The course then continues on to a series of special linear programming problems, such as optimal assignment, transportation, trans-shipment, network flow, minimal spanning tree, shortest path, PERT methods and traveling salesperson, each with its own algorithms and associated theory. Basic concepts of graph theory are normally introduced in conjunction with some of the preceding problems. If time permits, elementary aspects from decomposition theory, dynamic programming, integer programming, or nonlinear programming may be included.

It is difficult to find space in an introductory operations research course for even a small sampling of probability or stochastic models. If possible, it is better to include this material in a second course. Similarly, there is usually little time available to discuss game theory, except possibly for showing that two-person, zero-sum games are equivalent to a dual pair of linear programs. Game theory is probably best treated in a separate "topics" course.

Elementary Modeling Course

The following course on mathematical modeling and problem solving is intended for freshmen and sophomores with a solid preparation in high school mathematics. The primary objective is to provide lower-level students with a first-hand experience in forming their own mathematical models and discovering their own solution techniques. A secondary goal is to introduce some of the concepts from modern finite mathematics and illustrate their applications in the social sciences. The instructor might supplement these main themes with brief discussions of some important recent mathematical developments and indicate the current relevance of mathematics to contemporary science and policy making.

The course should maintain an open-minded and questioning approach to problem solving. Much of the class time should be devoted to student discussions of their models and how to improve them. Students should be asked to make formal oral and written expositions. Many of the topics covered are also suitable, with proper adjustments, for upper-level courses or for lower-level "mathematics appreciation" courses. (Readers interested in the latter courses should consult the 1981 Report of the CUPM Panel on Mathematics Appreciation, reprinted later in this volume.) Not all of the topics mentioned below can be covered in any one course, and frequent changes in course content are necessary to maintain the originality of problems.

No one current textbook appears appropriate for this course, although a simpler "prepackaged" version of this course could use the high-school-oriented text *You and Technology* with supplementary modules. The course described below is an example of how various sources can be assembled (as handouts or on library reserve) to form a modeling course, in this instance emphasizing modeling in the social sciences.

COURSE CONTENT

Overview and Patterns of Problem Solving. Introduction to the nature of modeling and problem solving. The role of science, engineering and social sciences in making and implementing new discoveries. The nature of applied mathematics and the interdisciplinary approach to problems. Illustrations of problems solved by quick insight rather than by involved analysis. Many books have chapters on modeling and problem solving; also see *Patterns of Problem Solving* by M. Rubinstein, Prentice-Hall, 1975, or "Foresight-Insight-Hindsight" by J. Frauenthal and T. Saaty, in *Modules in Applied Mathematics*, vol. 3 (W. Lucas, editor), Springer-Verlag.

Graph and Network Problems. A large variety of problems related to undirected and directed graphs and network flows can be assigned and discussed at the outset with no hint of any theory or technical terms. At a later stage, a lecture can be devoted to theory to develop a common vocabulary. The language and general approach of systems analysis can be developed. The four-color theorem can be discussed. References are *Applied Combinatorics*, by F. Roberts, Prentice-Hall, 1984, *Graphs as Mathematical Models* by G. Chartrand, Prindle, Weber and Schmidt, 1977, and *Applied Combinatorics* by A. Tucker, Wiley, 1980.

Some lecture time can be spent illustrating how graphs are applied: to simplify a complex problem, such as Instant Insanity (Chartrand, p. 125 or Tucker, p. 355), or the more difficult Rubik's Cube (*Scientific American,* March, 1981); for purely mathematical purposes, such as to prove Euler's formula $V - E + F = 2$ and use it in turn to prove the existence of exactly five regular polyhedra; or to examine R. Connelly's flexing (nonconvex) polyhedra (*Mathematical Intelligencer,* Vol. 1, No. 3, 1979). The analogy between transportation, fluid flow, electric and hydraulic networks can be illustrated (see G. Minty's article in *Discrete Mathematics and Its Applications* Proceedings of a Conference at Indiana University, ed. M. Thompson, 1977).

Enumeration Problems. (Tucker, 2nd ed., Chapter 5 or Roberts, Chapter 2.) Some practical uses can be covered briefly, e.g., to probability problems or the Pigeonhole Principle. Computational complexity and its application to hard-to-break codes can be discussed.

Value and Utility Theory. Expected utility versus expected value; St. Petersburg paradox; construction of a money versus utility curve: axioms for utility; assessing Coalitional Values (see module by W. Lucas and L. Billera in *Modules in Applied Mathematics,* vol. 2, W. Lucas, editor, Springer-Verlag).

Conflict Resolution. Some three-person cooperative game experiments and analysis; the Prisoner's Dilemma for two or more persons (H. Hamburger in *Journal of Math. Sociology* 3, 1973); illustrations of equilibrium concepts; two-person zero-sum games, e.g., batter versus pitcher (*Economics and the Competitive Process* by J. Case, NYU Press, 1979, p. 3; also see *The Game of Business* by John McDonald, Doubleday, 1975, Anchor paperback, 1977, and *Game Theory: A Nontechnical Introduction* by M. Davis, Basic Books, 1970).

A Discrete Optimization Problem and an Algorithm. Possible topics are the complete and optimal assignment problems (UMAP module 317 by D. Gale), or the marriage problem (D. Gale and L. Shapley, *American Mathematical Monthly* 69, 1962, p. 9).

Simulation. See chapters on simulation in many books and "Four-Way Stop or Traffic Light? An Illustration of the Modeling Process" by E. Packel (in *Modules in Applied Mathematics,* vol. 3, W. Lucas, editor, Springer-Verlag). Additional ideas from Inventory Theory, Scheduling Theory, Dynamic Programming, and Control Theory, e.g., lunar landing, can be included.

Projects and Mini-projects. At least one significant project type activity should be pursued over several weeks by the whole class by means of a sequence of graded exercises and class discussions. Some of the topics listed above can be treated in this mode. Other suitable topics are: the Apportionment Problem (Fair Representation by M. Balinski and H. Young, Yale Press); measuring power in Weighted Voting situations (W. Lucas in *Case Studies in Applied Mathematics* MAA, 1976); Cost Analysis (C. Clark in same *Case Studies* on harvesting fish or forests); some simple topics from statistics such as Asking Sensitive Questions, module by J. Maceli (in *Modules in Applied Mathematics,* vol. 2, W. Lucas, editor, Springer-Verlag); and Social Choice Theory and Voting (*Theory of Voting* by R. Farquharson, Yale, 1960).

In addition to the class project, teams of two or three students can spend a few weeks on a mini-project. Many of the topics above can be applied to a local practical problem. Scheduling, inventory and optimal allocations are good topics, as are gaming experiments, simulations and elementary statistical studies. More theoretical topics, ranging from walking versus running in the rain to designing the inside mechanism of the Rubik's Cube are also possible. Some attempt at discussing possible implementation of a mini-project result, e.g., with a campus administrator, is encouraged in order to show the practical difficulties of implementing mathematically optimal procedures.

Introductory Stochastic Processes

The purpose of this course is to introduce the student to the basic mathematical aspects of the theory of stochastic processes and its applications. This course can equally well be offered under such alternate titles as Applied Probability or Operations Research: Stochastic Models. Stochastic processes is a large and growing field. This course will lay background for further learning on the job or in graduate school.

The prerequisite for this course is at least the equivalent of a full course of post-calculus probability including the following topics: random variables, common univariate and multivariate distributions, moments, conditional probability, stochastic independence, conditional distributions and means, generating functions, and limit theorems. Such a course is fairly traditional now, but if most students have had just the integrated statistics and probability course suggested by the Statistics Subpanel, then the beginning of the stochastic processes course would have to be devoted to completing the needed probability background. It is also desirable for students to have some experience with basic matrix algebra and with using computer terminals.

The course should slight neither mathematical theory nor its applications. It is better to cover few topics with a full discussion of both theory and applications to survey theory alone or to cover only applications. The course emphasizes *problem solving* and develops an acquaintance with a variety of models that are widely used. Stochastic modeling and *problem formulation* are different activities that should be treated in a modeling course. If many students do not subsequently take a modeling course, then the instructor should consider allocating some time (assuming course time did not also have to be devoted to probability) to a module on stochastic modeling in business or government (see list of modules below) or to a real problem at the local college, e.g., modeling the demand for textbooks in the bookstore or utilization of campus parking spaces.

Computers should be used in this course in two ways:

- As a computational aid to perform, for example, matrix calculations needed in Markov chain theory; and
- As a simulation device to exhibit the behavior of random processes.

Understanding randomness is difficult for undergraduates and discussion of data accumulated in simulation studies can help overcome students' deterministic biases.

COURSE CONTENT

Bernoulli process; Markov chains (random walks, classification of states, limiting distributions); Poisson process (as limit of binomial process and as derived via axioms); Markov processes (transition functions and state probabilities, Kolmogorov equations, limiting probabilities, birth-death processes).

These basic topics have numerous applications that should be an essential feature of the course. In addition, some applied topics can be covered such as quality control, social and occupational mobility, Markovian decision processes, road traffic, reliability, queueing problems, population dynamics or inventory models. Instructors can find these and other applications in the many good texts on stochastic processes. Also see the modules and modeling texts listed at the end of this chapter.

Continuous Modeling

A primary goal of a continuous modeling course is to present the mathematical analysis involved in scientific modeling, as for example, the derivation of the heat equation. The course should also give an introduction to important applied mathematics topics, such as Fourier series, regular and singular perturbations, stability theory and tensor analysis. A few advanced topics can be chosen from boundary layer theory, nonlinear

waves and calculus of variations. The course should give a solid motivation for more advanced courses in these topics. A (non-original) paper on a topic of interest to the students serves the dual purpose of developing communication skills and introducing pedagogical flexibility.

A course on continuous modeling usually has as a prerequisite a course in differential equations, although the modeling can be taught concurrently or integrated in one course, using a book such as Martin Braun's *Differential Equations and Their Applications* (second edition), Springer-Verlag, 1978. Continuous modeling problems frequently involve concepts from natural sciences. In this case, it is important that either an appropriate background is required of students or the technical essentials are adequately introduced in the course.

The texts by Lin and Segal and by Haberman (see below) are well suited for this course. Selections from the two-volume Lin and Segal text can be used to provide a solid basis for physics and engineering modeling using both classical subjects, such as fluids, solids and heat transfer, and modern subjects, such as fields of biology. The text's broad coverage probably includes an introduction to an area of expertise of the instructor to which he or she can bring personal research insights.

A course which requires a little less sophistication can be designed around Haberman's book. This text's topics in population dynamics, oscillations, and traffic theory require less scientific background than topics in mechanics and mathematical biology, but still provide an excellent basis for modeling discussions. For example, population dynamics provide a good introduction to dynamical systems. Topics in regular and singular perturbation theory can be presented in the context of oscillations. Traffic theory provides a vehicle for introducing continuum mechanical modeling in which the processes are readily appreciated by students. Here the "microscopic" processes involve cars and drivers, and interesting models are obtained by car-following theory. Traffic flows also involve partial differential equations and shock waves.

References on Modeling

Modules

A. MODULE WRITING PROJECTS

Claremont Graduate School (Department of Mathematics):

- A Fractional Calculus Approach to a Simplified Air Pollution Model for Perturbation Analysis.

- Continuous-system Simulation Languages for DEC-10.
- Free Vibrations in the Inner Ear.
- Modeling of Stellar Interiors.
- Subsurface Areal Flow Through Porous Media.
- Variance Reduction for Monto Carlo Applications Involving Deep Penetration.
- Voting Games and Power Indices.

Mathematical Association of America's Committee on the Undergraduate Program in Mathematics Project, Case Studies in Applied Mathematics (designed especially for open-ended experiential teaching).
- Measuring Power in Weighted Voting Systems.
- A Model for Municipal Street Sweeping Operations.
- A Mathematical Model of Renewable Resource Conservation.
- Dynamics of Several-species Ecosystems.
- Population Mathematics.
- MacDonald's Work on Helminth Infections.
- Modeling Linear Systems by Frequency Response Methods.
- Network Analysis of Steam Generator Flow.
- Heat Transfer in Frozen Soil.

Mathematical Association of America Summer 1976 Module-writing Conference (at Cornell University Department of Operations Research):
- About sixty modules covering virtually all areas of application, such as biology, ecology, economics, energy, population dynamics, traffic flow, vibrating strings, and voting.
- Selected modules from this conference along with MAA applied mathematics case studies (ii) above were published by Springer-Verlag (New York, 1983) in four volumes, edited by William Lucas.

Rensselaer Polytechnic Institute (Department of Mathematical Sciences), published in *Case Studies in Mathematical Modeling*, by W. Boyce, Pitman, Boston, 1981:
- Herbicide Resistance.
- Elevator Systems.
- Traffic Flow.
- Shortest Paths in Networks.
- Computer Data Communication and Security.
- Semiconductor Crystal Growth.

State University of New York at Stony Brook (Department of Applied Mathematics and Statistics):
- A Model for Land Development.
- A Model for Waste Water Disposal, I and II.
- A Water Resource Planning Model.
- Man in Competition with the Spruce Budworm.
- Smallpox: When Should Routine Vaccination be Discontinued.

- Stochastic Models for the Allocation of Fire Companies.

B. MODULES DEVELOPED BY INDIVIDUALS
- Undergraduate Mathematics Application Project (UMAP): UMAP has several hundred modules covering all areas of application. Selected modules appear in the *UMAP Journal* (four issues a year), published by Birkhauser-Boston. UMAP catalogue available by writing to: UMAP, Educational Development Center, 55 Chapel Street, Newton, MA 02160.

C. PROCEEDINGS OF MODELING CONFERENCES
1. Discrete Mathematics and Its Applications, Proceedings of a Conference at Indiana University, ed. M. Thompson, 1976.
2. Mathematical Models in the Undergraduate Curriculum, Proceedings of Conference at Indiana University, ed. D. Maki and M. Thompson, 1975.
3. Proceedings of Summer Seminar on Applied Mathematics, ed. M. Thompson, Indiana University, 1978.
4. Mathematical Models for Environmental Problems, Proceedings of the International Conference at the University of Southampton, 1976.
5. Proceedings of Conference on Environmental Modeling and Simulation, Environmental Protection Agency, 1976.
6. Proceedings of a Conference on the Application of Undergraduate Mathematics in the Engineering, Life, Managerial and Social Sciences, ed. P. Knopp and G. Meyer, Georgia Institute of Technology, 1973.
7. Proceedings of the Pittsburgh Conferences on Modeling and Simulations, Vols. 1-9 (1969-78), Instrument Society of America.
8. Proceedings of the Summer Conference for College Teachers on Applied Mathematics, University of Missouri-Rolla, 1971.
9. Information Linkage Between Applied Mathematics and Industry, ed. P. Wang, Academic Press, 1976.

Articles on Teaching Modeling
1. J. Agnew and M. Keener, A Case-study Course in Applied Mathematics Using Regional Industries, *American Mathematical Monthly* 87 (1980).
2. R. Barnes, Applied Mathematics: An Introduction Via Models, *American Mathematical Monthly* 84 (1977).
3. C. Beaumont and R. Wieser, Co-operative Programmes in Mathematical Sciences at the University of Waterloo, *Journal of Co-operative Education* 11 (1975).

4. J. Becker, R. Borrelli, and C. Coleman, *Models for Applied Analysis,* Harvey Mudd College, 1976 and revised annually.

5. R. Borrelli and J. Spanier, The Mathematics Clinic: A Review of Its First Seven Years, *UMAP Journal* 2 (1981).

6. R. Borton, *Mathematical Clinic Handbook,* Claremont Graduate School, 1979.

7. J. Brookshear, A Modeling Problem for the Classroom, *American Mathematical Monthly* 85 (1978).

8. E. Clark, *How To Select a Clinic Project,* Harvey Mudd College, 1975.

9. C. Hall, Industrial Mathematics: A Course in Realism, *American Mathematical Monthly* 82 (1975).

10. L. Handa, *Mathematics Clinic Student Handbook: A Primer for Project Work,* Harvey Mudd College, 1979.

11. J. Hachigian, Applied Mathematics in a Liberal Arts Context, *American Mathematical Monthly* 85 (1978).

12. E. Rodin, Modular Applied Mathematics for Beginning Students, *American Mathematical Monthly* 84 (1977).

13. R. Rubin, Model Formulation Using Intermediate Systems, *American Mathematical Monthly* 86 (1979).

14. M. Seven and T. Zagar, *The Engineering Clinic Guidebook,* Harvey Mudd College, 1975.

15. D. Smith, A Seminar in Mathematical Modelbuilding, *American Mathematical Monthly* 86 (1979).

16. J. Spanier, The Mathematics Clinic: An Innovative Approach to Realism Within an Academic Environment, *American Mathematical Monthly* 83 (1976).

Books on Mathematical Modeling

For further references, see Applications section of *A Basic Library List,* Mathematical Association of America, 1976.

A. GENERAL MODELING

1. J. Andrew and R. McLone, ed., *Mathematical Modeling,* Butterworth, 1976.

2. R. Aris, *Mathematical Modeling Techniques,* Pitman, 1978.

3. E. Beltrami, *Mathematics for Dynamic Modeling,* Academic Press, 1987.

4. E. Bender, *An Introduction to Mathematical Modeling,* Wiley, 1978.

5. G. Carrier, *Topics in Applied Mathematics,* Vol. I and II, MAA summer seminar lecture notes, Mathematical Association of America, 1966.

6. C. Coffman and G. Fix, ed., *Constructive Approaches to Mathematical Models,* Academic Press, 1980.

7. R. DiPrima, ed., *Modern Modeling of Continuous Phenomena,* American Mathematical Society, 1977.

8. C. Dym and E. Ivey, *Principles of Mathematical Modeling,* Academic Press, 1980.

9. B. Friedman, *Lectures on Applications-oriented Mathematics,* Holden-Day, 1969.

10. F. Giordano and M. Weir, *A First Course in Mathematical Modeling,* Brooks/Cole, 1985.

11. P. Lancaster, *Mathematics Models of the Real World,* Prentice Hall, 1976.

12. D. Maki and M. Thompson, *Mathematical Models and Applications,* Prentice Hall, 1976.

13. F. Roberts, *Discrete Mathematical Models,* Prentice Hall, 1976.

14. T. Saaty, *Thinking with Models,* AAAS Study Guides on Contemporary Problems No. 9, 1974.

B. MODELING IN VARIOUS DISCIPLINES

Mathematical modeling is such an integral part of physics and engineering that any text in these fields is implicitly a mathematical modeling book.

1. P. Abell, *Model Building in Sociology,* Shocken, 1971.

2. R. Aggarwal and I. Khera, *Management Science Cases and Applications,* Holden-Day, 1979.

3. R. Atkinson, et al., *Introduction to Mathematical Learning Theory,* Krieger Publishing, 1965.

4. D. Bartholomew, *Stochastic Models for Social Processes,* Wiley, 1973.

5. M. Bartlett, *Stochastic Population Models,* Methuen, 1960.

6. R. Barton, *A Primer on Simulation and Gaming,* Prentice Hall, 1970.

7. S. Brams, *Game Theory and Politics,* The Free Press, 1975.

8. C. Clark, *Mathematical Bioeconomics,* Wiley, 1976.

9. J. Coleman, *Introduction to Mathematical Sociology,* Free Press, 1964.

10. P. Fishburn, *The Theory of Social Choice,* Princeton University Press, 1973.

11. J. Frauenthal, *Introduction to Population Modeling,* UMAP Monograph, 1979.

12. H. Gold, *Mathematical Modeling of Biological Systems,* Wiley, 1977.

13. S. Goldberg, *Some Illustrative Examples of the Use of Undergraduate Mathematics in the Social Sciences,* Mathematical Association of America, CUPM Report, 1977.

14. M. Gross, *Mathematical Models in Linguistics*, Prentice Hall, 1972.

15. R. Haberman, *Mathematical Models, Mechanical Vibrations, Population Dynamics and Traffic Flow*, Prentice Hall, 1977.

16. F. Hoppensteadt, *Mathematical Theories of Populations: Demographics and Epidemics*, SIAM, 1976.

17. J. Kemeny and L. Snell, *Mathematical Models in the Social Sciences*, MIT Press, 1973.

18. C. Lave and J. March, *An Introduction to Models in the Social Sciences*, Harper and Row, 1975.

19. C. Lin and L. Segal, *Mathematics Applied to Deterministic Problems in the Natural Sciences*, Macmillan, 1974.

20. D. Ludwig, *Stochastic Population Theories*, Springer, 1974.

21. J. Maynard-Smith, *Models in Ecology*, Cambridge University Press, 1974.

22. B. Noble, *Applications of Undergraduate Mathematics to Engineering*, Mathematical Association of America, 1976.

23. M. Olinik, *An Introduction to Mathematical Models in the Social and Life Sciences*, Addison Wesley, 1978.

24. E. Pielou, *Mathematical Ecology*, Wiley, 1977.

25. H. Pollard, *Mathematical Introduction to Celestial Mechanics*, Mathematical Association of America, 1977.

26. J. Pollard, *Mathematical Models for the Growth of Human Populations*, Cambridge University Press, 1973.

27. D. Riggs, *The Mathematical Approach to Physiological Problems*, Macmillan, 1979.

28. T. Saaty, *Topics in Behavioral Mathematics*, MAA summer seminar lecture notes, Mathematical Association of America, 1973.

29. H. Scarf, et al., *Notes on Lectures on Mathematics in the Behavioral Sciences*, MAA summer seminar lecture notes, Mathematical Association of America, 1973.

30. C. von Lanzenauer, *Cases in Operations Research*, Holden Day, 1975.

31. H. Williams, *Model Building in Mathematical Programming*, Wiley, 1978.

Subpanel Members

WILLIAM LUCAS, CHAIR, Cornell University.

ROBERT BORRELLI, Harvey Mudd College.

RALPH DISNEY, VPI & SU.

DONALD DREW, RPI.

SAM GOLDBERG, Oberlin College.

Statistics

This chapter contains the report of the Subpanel on Statistics of the CUPM Panel on a General Mathematical Sciences Program, reprinted with minor changes from Chapter VI of the 1981 CUPM report entitled RECOMMENDATIONS FOR A GENERAL MATHEMATICAL SCIENCES PROGRAM.

Introductory Course

Statistics is the methodological field of science that deals with collecting data, organizing and summarizing data, and drawing conclusions from data. Although statistics makes essential use of mathematical tools, especially probability theory, it is a misrepresentation of statistics to present it as essentially a subfield of mathematics.

The Statistics Subpanel believes that an introductory course in probability and statistics should concentrate on data and on skills and mathematical tools motivated by the problems of collecting and analyzing data. The traditional undergraduate course in statistical theory has little contact with statistics as it is practiced and is not a suitable introduction to the subject. Such a course gives little attention to data collection, to analysis of data by simple graphical techniques, and to checking assumptions such as normality.

The field of statistics has grown rapidly in applied areas such as robustness, exploratory data analysis, and use of computers. Some of this new knowledge should appear in a first course. It is now inexcusable to present the two-sample t-test for means and the F-test for variances as equally legitimate when a large literature demonstrates that the latter is so sensitive to non-normality as to be of little practical value, while the former (at least for equal sample sizes) is very robust (e.g., see Pearson and Please, *Biometrika* 62 (1975), pp. 223-241, for an effective demonstration). However, the Statistics Subpanel does not believe that a course in "exploratory data analysis" is a suitable introduction to statistics, nor does it advocate replacing (say) least squares regression by a more robust procedure in a first course. But it does think that new knowledge renders a course devoted solely to the theory of classical parametric procedures out of date.

While the Statistics Subpanel prefers a two-semester introductory sequence in probability and statistics, enrollment data shows that most students take only a single course in this area. The course proposed below gives students a representative introduction to both the data-oriented nature of statistics and the mathematical concepts underlying statistics. These broad objectives raise several issues that require preliminary comment. One year of calculus is assumed for this course. The course should use Minitab or a similar interactive statistical package.

The Place of Probability

Probability is an essential tool in several areas of the mathematical sciences. It is not possible to compress a responsible introduction to probability and coverage of statistics into a single course. The Statistics Subpanel therefore recommends that probability topics be divided between the courses on probability and statistics, discrete methods, and modeling/operations research as follows:

- *Probability and statistics course:* Axioms and basic properties; random variables; univariate probability functions and density functions; moments; standard distributions; Laws of Large Numbers and Central Limit Theorem.
- *Discrete methods course:* Combinatorial enumeration problems in discrete probability.
- *Modeling/operations research course:* Conditional probability and several-stage models; stochastic processes.

This division is natural in the sense that the respective parts of probability are motivated by and applied to the primary concerns of these courses.

Alternative Arrangements

The subpanel is convinced that two semesters are required for a firm introduction to both probability and statistics. Many institutions now offer such a two-semester sequence in which probability is followed by statistics. The subpanel prefers this structure. In this sequence the statistics course should be revised to incorporate the topics and flavor of the data analysis section of the proposed unified course. With probability first, added material in statistics can also be covered, such as Neyman-Pearson theory, distribution-free tests, robust procedures, and linear models.

Institutions will vary considerably in their choice of material for this statistics course, but the subpanel reiterates its conviction that the traditional "theory-only" statistics course is not a wise choice. If experience shows that many students drop out in the middle of a two-course sequence, the unified course outlined below should be adopted, followed by one of the elective courses suggested in Section 3 of this chapter.

Instructor Preparation

Since the recommended outline is motivated by data and shaped by the modern practice of statistics, many mathematically trained instructors will be less prepared to teach this course than a traditional statistical theory course. Growing interest in "applied" statistics has, of course, led many instructors to broaden their knowledge. Some background reading is provided for others who wish to do so. The publications listed here contain material that can be incorporated in the recommended course, but none is suitable as a course text. In order of ascending level:

1. Tanur, Judith, *et al.*, eds., *Statistics: A Guide to the Unknown*, Second Edition, Holden-Day, 1978.

     An elementary volume describing important applications of statistics and probability in many fields of endeavor.

2. Moore, David, *Statistics: Concepts and Controversies*, W.H. Freeman, 1979.

     A paperback with good material on data collection, statistical common sense, appealing examples, and the logic of inference.

3. Freedman, David; Pisani, Robert; Purves, Roger, *Statistics*, W.W. Norton, 1978.

     A careful introduction to elementary statistics written with conceptual richness, attention to the real world, and awareness of the treachery of data.

4. Mosteller, Frederick and Tukey, John, *Data Analysis and Regression*, Addison-Wesley, 1977.

     Good ideas on exploratory data analysis, robustness and regression.

5. Box, George; Hunter, William; Hunter, J. Stuart, *Statistics for Experimenters*, Wiley, 1978.

     Applied statistics explained by experienced practical statisticians. Some specialized material, but much of the book will repay careful reading.

6. Efron, Bradley, "Computers and the Theory of Statistics: Thinking the Unthinkable," *SIAM Review*, October, 1979.

     A superb article on some new directions in statistics, written for mathematicians who are not statisticians.

Course Outline

I. Data   (about 2 weeks)

- *Random sampling.* Using a table of random digits; simple random samples, experience with sampling variability of sample proportions and means; stratified samples as a means of reducing variability.
- *Experimental design.* Why experiment; motivation for statistical design when field conditions for living subjects are present; the basic ideas of control and randomization (matching, blocking) to reduce variability.

COMMENTS: Data collection is an important part of statistics. It meets practical needs (see Moore) and justifies the assumptions made in analyzing data (see Box, Hunter and Hunter). Experience with variability helps motivate probability and the difficult idea of a sampling distribution. Students should see for themselves the results of repeated random sampling from the same population and the variability of data in simple experiments such as comparing 3-minute performance of egg timers (see W.G. Hunter, *American Statistician*, 1977, pp. 12-17, for suggestions).

II. Organizing and Describing Data

                                              (about 2 weeks)

- *Tables and graphs.* Frequency tables and histograms; bivariate frequency tables and the misleading effects of too much aggregation; standard line and bar graphs and their abuses; box plots; spotting outliers in data.
- *Univariate descriptive statistics.* Mean, median and percentiles; variance and standard deviation; a few more robust statistics such as the trimmed mean.
- *Bivariate descriptive statistics.* Correlation; fitting lines by least squares. If computer resources permit, least-square fitting need not be restricted to lines.

COMMENTS: In addition to simple skills, students must be trained to look at data and be aware of pitfalls. Freedman, Pisani and Purves have much good material on this subject, such as the perils of aggregation (pp. 12-15). The impressive effect on a correlation of keypunching 7.314 as 731.4 should be pointed out. Simple plots are a powerful tool and should be stressed throughout the course as part of good practice.

III. Probability   (about 4 weeks)

- *General probability.* Motivation; axioms and basic rules, independence.
- *Random variables.* Univariate density and probability functions; moments; Law of Large Numbers.

- *Standard distributions.* Binomial, Poisson, exponential, normal; Central Limit Theorem (without proof).
- *More experience with randomness.* Use in computer simulation to illustrate Law of Large Numbers and Central Limit Theorem.

COMMENTS: Probability must unavoidably be pressed in a unified course that includes data analysis. Instructors should repeatedly ask "What probability do I need for basic statistics?" and "What can the students learn within about four weeks?" It is certainly the case that combinatorics, moment generating functions, and continuous joint distributions must be omitted. Some instructors may be able to cover conditional probability and Bayes' theorem in addition to the outline material.

IV. Statistical Inference   (about 6 weeks)

- *Statistics vs. probability.* The idea of a sampling distribution; properties of a random sample, e.g., it is normal for normal populations; the Central Limit Theorem.
- *Tests of significance.* Reasoning involved in alpha-level testing and use of P-values to assess evidence against a null hypothesis; cover one- and two-sample normal theory tests and (optional) chi-square tests for categorical data. Comment on robustness, checking assumptions, and the role of design (Part I) in justifying assumptions.
- *Point estimation methods.* Method of moments; maximum likelihood; least squares; unbiasedness and consistency.
- *Confidence intervals.* Importance of error estimate with point estimator; measure of size of effect in a test of significance.
- *Inference in simple linear regression.*

COMMENTS: A firm grasp of statistical reasoning is more important than coverage of a few additional specific procedures. For much useful material on statistical reasoning such as use of the "empirical rule" to assess normality, see Box, Hunter and Hunter. *Don't* just say, "We assume the sample consists of iid normal random variables." Applied statisticians favor P-values over fixed alpha tests; a comparative discussion of this issue appears in Moore.

RECOMMENDED TEXTS

The Subpanel is not aware of a text at the post-calculus level that fits the recommended outline closely. Instructors should seriously consider adopting a good post-calculus statistical methods text rather than a theoretical statistics text. A methods text is more likely to have examples and problems which have the ring of

truth. Moreover, most instructors will find it easy to supplement a methods text with mathematical material and problems familiar from previous teaching. It is much harder to supply motivation and realistic problems, and it is psychologically difficult for both the teacher and student to skip much of the probability in a mathematical statistics text.

The following books are possible texts or reference material for the course described above. All of these have essentially the same shortcoming of being too "unmathematical." The appropriate combination of level of sophistication and content is not now available under a single cover. The class of books below fall in the "intermediate" level between an elementary statistics course and a first course in mathematical statistics.

1. Box, George; Hunter, William; Hunter, J. Stuart, *Statistics for Experimenters: An Introduction to Design, Data Analysis, and Model Building,* John Wiley & Sons, New York, 1978.
2. Moore, David and McCabe, George, *Introduction to the Practice of Statistics,* Freeman, San Francisco, 1989.
3. Ott, Lyman, *An Introduction to Statistical Methods and Data Analysis, Second Edition,* Duxbury, Boston, 1984.
4. Neter, John; Wasserman, William; Whitmore, G.A., *Applied Statistics,* Allyn and Bacon, Boston, 1978.

Additional Courses

Probability and Statistical Theory

CONTENT: Distribution functions; moment and probability generating functions; joint, marginal and conditional distributions; correlations; distributions of functions of random variables; Chebyschev's inequality; convergence in probability; limiting distributions; power test and likelihood ratio tests; introduction to Bayesian and nonparametric statistics; additional regression topics.

COMMENT: This course is designed to complete the traditional probability-then-statistics sequence. Since the students have already completed a semester of study, they should be capable of tackling a good text on mathematical statistics such as the one by DeGroot or by Hogg and Craig. The book by Bickel and Doksum is a little more difficult than the other two, and the teacher would have to supplement it with the topics in probability.

TEXTS:

1. Mendenhall, William; Schaeffer, Richard; Wackerly, Dennis, *Mathematical Statistics with Applications, Second Edition,* Duxbury, Boston, 1981.
2. Larsen, Richard and Marx, Morris, *An Introduction to Probability and its Applications,* Prentice-Hall, Englewood Cliffs, N. Jers., 1985.
3. DeGroot, Morris M., *Probability and Statistics,* Addison-Wesley, Reading, Mass., 1975.
4. Hogg, Robert and Craig, Allen, *Introduction to Mathematical Statistics,* Macmillan, New York, 1978.

Applied Statistics

CONTENT: This course uses statistical packages to analyze data sets. Topics include linear and multiple regression; nonlinear regression; analysis of variance; random, fixed and mixed models; expected mean squares; pooling, modifications under relaxed assumptions; multiple comparisons; variance of estimators; analysis of covariance.

COMMENT: The new introductory course will probably attract more students from other fields than the traditional probability-then-statistics course. This course is an excellent follow-up for such non-mathematical sciences students. Its topics are among the more widely used statistical tools. Students should be expected to use a statistical computing package such as Minitab of SPSS for many of the analyses. The book by Miller and Wichern is a possible text for this course.

TEXTS:

1. Miller, Robert and Wichern, Dean, *Intermediate Business Statistics,* Holt, Rinehart and Winston, New York, 1977.
2. Neter, John; Wasserman, William; Kutner, Michael, *Applied Linear Statistical Models, Second Edition,* Irwin, 1985.
3. Morrison, Donald, *Applied Linear Statistical Methods,* Prentice-Hall, Englewood Cliffs, N. Jers., 1983.

Probability and Stochastic Processes

CONTENT: Combinatorics; conditional probability and independence; Bayes theorem; joint, marginal and conditional distributions; distribution functions; distributions of functions of random variables; probability and moment generating functions; Chebyschev's inequality; convergence in probability; convergence in distribution; random walks; Markov chains; introduction to continuous-time stochastic processes.

COMMENT: This is a fairly standard course and a number of texts are available. The book by Feller is a classic but covers only discrete probability. The book by Olkin, Gleser and Derman is at a slightly lower level and is more "applied" but will require the instructor to provide some supplementary materials. The book by Chung is excellent but must be read with a "grain of salt." The book by Breiman is also excellent but expects much of its reader. A new book by Johnson and Kotz also looks interesting but is restricted to discrete probability. The books by Chung, Feller and Breiman are difficult for the average student.

TEXTS:

1. Olkin, Ingram; Gleser, Leon J.; Derman, Cyrus, *Probability Models and Applications,* Macmillan, New York, 1980.
2. Larsen, Richard and Marx, Morris, *An Introduction to Probability and its Applications,* Prentice-Hall, Englewood Cliffs, N.J., 1985.
3. Ross, Sheldon, *A First Course in Probability, Second Edition,* Macmillan, New York, 1984.
4. Chung, Kai Lai, *Elementary Probability Theory with Stochastic Processes,* Springer-Verlag, New York, 1974.
5. Feller, William, *An Introduction to Probability Theory and Its Applications, Volume I,* John Wiley & Sons, New York, 1950.

Preparation for Graduate Study

There are a large number of career opportunities for statisticians in industry, government and teaching. For example as of 1977, the Federal Government employed over 3500 statisticians, plus 3500 statistical assistants and numerous other employees performing statistical duties but classified in different job series. A recent report by the U.S. Labor Department, reprinted in the *New York Times* National Recruitment Survey, predicts an increase of 35% in the demand for statisticians during the 1980's. This compares to a predicted increase of 9% for mathematicians and 30% for computer specialists.

Preparation for a career in statistics usually involves graduate study. An undergraduate major in statistics, computer science, or mathematical sciences is the recommended preparation for graduate study in statistics. It is desirable for such a major to include solid courses in matrix theory and real analysis, in addition to courses in probability and statistics. Most statistics graduate programs require matrix algebra and real analysis for fully matriculated admission. Either one of the sample programs in the report of the General Mathematical Sciences Panel in the first chapter would be adequate preparation for graduate study in statistics. However, major A is preferable to major B, and both should include at least one follow-on elective in probability and statistics.

In addition to courses in the mathematical sciences, a student preparing for graduate study in statistics should:

- Study in depth some subject where statistics is an important tool (physics, chemistry, economics, psychology, ...). In fact, a double major should be considered.
- Take as many courses as possible which are designed to enhance his or her communication skills. Statisticians in industry and government are often called upon to provide written reports and critiques; consulting requires clear oral communications.

A detailed discussion of preparation for a statistical career in industry can be found in [1] A similar report, [2], discusses preparation for a career in government.

1. Preparing Statisticians for Careers in Industry: Report of the ASA Section on Statistical Education. *The American Statistician,* 1980, pp. 65-80.
2. Preparing Statisticians for Careers in Government: Report of the ASA Section on Statistics in Government. Paper presented at the American Statistical Association meeting in August, 1980.

Panel Members:

RICHARD ALO, CHAIR, Lamar University.
RICHARD KLEBER, St. Olaf College.
DAVID MOORE, Purdue University.
MIKE PERRY, Appalachian State University.
TIM ROBERTSON, University of Iowa.

Discrete Mathematics

In the early 1980's, as computer science enrollments ballooned on campuses across the country, the Mathematical Association of America established an ad hoc Committee on Discrete Mathematics to help provide leadership to the rapidly expanding efforts to create a course in discrete mathematics that would meet the needs of computer science and at the same time fit well into the traditional mathematics program. This committee issued a report in 1986 that conveyed their own recommendations together with an appendix that included reports from six experimental projects supported by the Alfred P. Sloan Foundation. This chapter contains the report of the Committee, without the appendix, preceded by a new preface prepared by Committee Chair Martha Siegel.

1989 Preface

In the years since 1986 when the Committee on Discrete Mathematics in the First Two Years published its report, there have been many changes in the attitudes of mathematics departments toward curricular change. The Committee had found that faculty in disciplines that required calculus were quite supportive of proposals to introduce more discrete mathematics into the first two years. They frequently complained about the state of calculus and encouraged us to get our house in better order. Many mathematicians also expressed dissatisfaction with the calculus sequence. The threat of replacing some of the traditional calculus material with discrete topics certainly helped to turn attention to the teaching of calculus. This movement toward a "calculus for a new century" is exciting and timely.

It is disappointing, however, that there seem to be only a few attempts to incorporate any significant discrete mathematics into the revision of the curriculum of the first two years. The discrete mathematics course seems to be established in most schools as a separate entity. It is encouraging that the National Council of Teachers of Mathematics has established a Task Force on Discrete Mathematics to help teachers implement curriculum standards for the inclusion of discrete mathematics in the schools.

Many new textbooks for the standard (usually one semester) discrete mathematics courses for the freshman or sophomore student have appeared or are in press. Publishers seem to find the market troublesome because there is no consensus as to the exact content of the course. Those suggestions offered by the Committee as to topics that should be considered for inclusion may have been loosely followed, but level and attitude of books vary widely. A bibliography which is current as of the end of 1988 appears in a report on discrete mathematics edited by Anthony Ralston (to appear in the MAA Notes Series, 1989), as do final reports of the Sloan Foundation funded discrete mathematics projects.

New freshman-sophomore textbooks continue to appear. For the immediate future, it seems as if a one- or two-semester discrete mathematics course, independent of (but at the same level as) calculus will be the typical one. Recent advances such as ISETL, a computer language for the teaching of discrete mathematics (*Learning Discrete Mathematics with ISETL*, by Nancy Baxter, Ed Dubinsky, and Gary Levin), and the *True BASIC Discrete Mathematics* package(by John Kemeny and Thomas Kurtz, Addison-Wesley) may affect how the course evolves. Those few efforts to incorporate discrete mathematics into the calculus and the courseware that are being developed merit our attention for the future.

Who teaches discrete mathematics? Most mathematics departments have a course, though sometimes only on the junior-senior level. Sometimes the elementary course is offered in the computer science department or in engineering. The 1985-86 CBMS survey indicates that more than 40% of all institutions require discrete mathematics for computer science majors. Of universities and four-year colleges, about 60% require discrete mathematics or discrete structures. It is also not uncommon that mathematics majors are required to take some discrete mathematics.

The most recent accreditation standards issued by the Computer Sciences Accreditation Board include a discrete mathematics component. The most recent report of the ACM Task Force on the Core of Computer Science defines nine areas as the core of computer science (*Communications of ACM*, Jan. 1989, 32:1). In all but a few areas, discrete mathematics topics are listed as support areas. For algorithms and data structures, for example, students should be familiar with graph theory, recursive functions, recurrence relations, combinatorics, induction, predicate and temporal logic, among other things. Boolean algebra and coding theory are considered part of the architecture component. Students certainly would need discrete mathematics as a prerequisite for many of the computer core courses. Graph theory, logic, and algebra appear in a significant number of necessary support areas.

Although the Committee on Discrete Mathematics in the First Two Years was dissolved after its 1986 report was issued, concerns of the Committee have been incorporated into the mission of the MAA Committee on Calculus Revision and the First Two Years (CRAFTY). Aside from concentration on the calculus initiatives across the country, CRAFTY is interested in continuing the effort of the earlier group to see discrete mathematics become part of the typical freshman-sophomore curriculum in any of the mathematical sciences. The goal is to increase the effectiveness of the curriculum in serving other disciplines while providing enough excitement and challenge to attract talented undergraduates to major in mathematics.

Martha J. Siegel
Towson State University
March, 1989

Introduction and History

The Committee on Discrete Mathematics in the First Two Years was established in the spring of 1983 for the purpose of continuing the work begun at the Williams College Conference held in the summer of 1982. That conference brought to a forum the issue of revising the college curriculum to reflect the needs of modern programs and the students in them. Anthony Ralston and Gail Young brought together 29 scientists (24 of whom were mathematicians) from both industry and academe to discuss the possible restructuring of the first two years of college mathematics. Although the growing importance of computer science majors as an audience for undergraduate mathematics was an important motivation for the Williams Conference, the conference concerned itself quite broadly with the need to revise the first two years of the mathematics curriculum for everyone—mathematics majors, physical science and engineering majors, social and management science majors as well as computer science majors. The papers presented and discussed at the conference, and collected in *The Future of College Mathematics* [35], reflect this breadth of view.

The word used to describe what was needed was "discrete" mathematics. Most of us knew what that meant approximately and respected the content as good mathematics. To illustrate the discrete mathematics topics

that might be considered for an elementary course, two workshop groups at the Williams Conference produced (in a very short time) a fairly remarkable set of two course sequences:

1. A two year sequence of independent courses, one in discrete mathematics and one in a streamlined calculus, and

2. A two year integrated course in discrete and continuous mathematics (calculus) in a modular form for service to many disciplines.

These course outlines were admittedly tentative and needed refinement and testing. At the same time, the CUPM-CTUM Subcommittee on Service Courses had been examining the traditional service course offerings of the first two years. The syllabi of these courses, in which many freshman and sophomores are required to enroll, are studied periodically for their relevancy. Finite mathematics, linear algebra, statistics, and calculus are considered to be essential to many majors, but with the importance of the computer, the Subcommittee on Service Courses concluded that even the mathematics majors need mathematics of a new variety, not only so they can take computer science courses, but also so they can work on contemporary problems in mathematics.

At that time, there were few or no textbooks or examples of such courses for the community to share. At the suggestion of the Subcommittee on Service Courses, the MAA agreed to help to develop the Williams courses further through the Committee on Undergraduate Program in Mathematics (CUPM) and the Committee on the Teaching of Mathematics (CTUM), standing committees of the Association. That led to the establishment of the committee responsible for this report. Funds for the effort were secured from the Sloan Foundation. The members of the committee were chosen especially to reflect the communities who would eventually be most affected by any changes in the traditional mathematics curriculum.

In addition to the development of course outlines and plans for their implementation, the committee was also involved in the observation of a set of experimental projects which also were begun as a result of the Williams Conference and the interest of the Sloan Foundation. After the conference, the Sloan Foundation solicited about thirty proposals for courses which would approximate the syllabi suggested by the workshop participants. The call for proposals particularly mentioned the need for the development of text material and classroom testing and emphasized the hope that some schools would make the effort to try the integrated curriculum. Six schools were ultimately chosen:

- Colby College, Waterville, Maine.
- University of Delaware, Newark, Delaware.
- University of Denver, Denver, Colorado.
- Florida State University, Tallahassee, Florida.
- Montclair State College, Montclair, New Jersey.
- St. Olaf College, Northfield, Minnesota.

The committee, together with some of the committee which chose the proposals to be funded (Don Bushaw, Steve Maurer, Tony Ralston, Alan Tucker, and Gail Young), monitored their progress for the two year period of funding, ending in August 1985. (Complete reports of these funded projects will appear in an MAA Notes volume on discrete mathematics to be published in 1989.)

Summary of Recommendations

1. Discrete mathematics should be part of the first two years of the standard mathematics curriculum at all colleges and universities.
2. Discrete mathematics should be taught at the intellectual level of calculus.
3. Discrete mathematics courses should be one-year courses which may be taken independently of the calculus.
4. The primary themes of discrete mathematics courses should be the notions of proof, recursion, induction, modeling, and algorithmic thinking.
5. The topics to be covered are less important than the acquisition of mathematical maturity and of skills in using abstraction and generalization.
6. Discrete mathematics should be distinguished from finite mathematics, which as it is now most often taught, might be characterized as baby linear algebra and some other topics for students not in the "hard" sciences.
7. Discrete mathematics should be taught by mathematicians.
8. All students in the sciences and engineering should be required to take some discrete mathematics as undergraduates. Mathematics majors should be required to take at least one course in discrete mathematics.
9. Serious attention should be paid to the teaching of the calculus. Integration of discrete methods with the calculus and the use of symbolic manipulators should be considered.
10. Secondary schools should introduce many ideas of discrete mathematics into the curriculum to help students improve their problem-solving skills and prepare them for college mathematics.

General Discussion

In its final report, the committee has decided to present two course outlines for elementary mainstream discrete mathematics courses. Our unanimous preference is for a one-year course, at the level of the calculus but independent of it. It is designed to serve as a service course for computer science majors and others and as a possible requirement for mathematics majors. The Committee on the Undergraduate Program in Mathematics (CUPM) has endorsed the recommendation that every mathematics major take a course in discrete mathematics, and has agreed that the year course the committee recommends is a suitable one for mathematics majors. It is expected that the course will be taken by freshmen or sophomores majoring in computer science so that they can apply the material in the first and second year courses in their major. The ACM recommendations [21, 22] for the first year computer science course presume, if not specific topics, then certainly the level of maturity in mathematical thought which students taking the discrete mathematics course might be expected to have attained. Hence, the Committee recommends that the course be taken simultaneously with the first computer science course. The Committee understands that at some schools the first computer science course may be preceded by a course strictly concerned with programming. At the very least, the Committee expects that the discrete mathematics course will be a prerequisite to upper-level computing courses. For this reason, the Committee has tried to isolate those mathematical concepts that are used in computer science courses. The usual sequence of these courses might determine what should be taught in the corresponding mathematics courses.

In addition to the Committee's concern for computer science majors, there is a high expectation that mathematics majors and those in most physical science and engineering fields will benefit from the topics and the problem-solving strategies introduced in this discrete mathematics course. Subjects like combinatorics, logic, algebraic structures, graphs, and network flows should be very useful to these students. In addition, methods of proof, mathematical induction, techniques for reducing complex problems to simpler (previously solved) problems, and the development of algorithms are tools to enhance the mathematical maturity of all. Furthermore, students in these scientific and mathematically-oriented fields will want to take computer science courses, and will need some of the same mathematical preparation that the computer science major needs.

Thus, the Committee has agreed to recommend that

the course be part of the regular mathematics sequence in the first two years for all students in mathematically-related majors. Our contacts with physicists and engineers reinforce the idea that their students will need this material, but, of course, there is the concern that calculus will be short-changed.

The Committee will make several suggestions regarding the calculus, but individual institutions will best understand their own needs in this regard. We do not recommend that the third semester of calculus be cut from a standard curriculum. Serious students in mathematical sciences, engineering, and physical sciences need to know multivariable calculus. Many in the mathematical community recognize that the content of the calculus should be updated to acknowledge the use of numerical methods and computers, and promising initiatives along this line are being taken. Engineers have been especially anxious for this change. John Schmeelk surveyed 34 schools and compiled suggestions for revising the standard calculus. (Schmeelk's survey was included in the appendix to the original report of the Committee.) At some of the Sloan-funded schools and others, there have been attempts to revise the calculus to incorporate some discrete methods and to use the power of the symbolic manipulator packages. We describe these attempts later in the report. (A complete report on the Sloan-funded projects will be published in the MAA Notes Series in 1989.)

There is, inherent in our proposal, the possibility that some students may be required to take five semesters of mathematics in the first two years—a year of discrete mathematics and the three semesters of calculus. But, there is no reason why students cannot be allowed to take one of the five in the junior year. We point out that some linear algebra is included in the year of discrete mathematics. Additionally, the use of computers via the new and powerful symbolic manipulation packages may reduce the amount of time needed for the traditional calculus sequence.

A one-semester discrete mathematics course will be described in the appendix to this report as a concession to the political realities in many institutions. It has become obvious to the Committee over the last two years that at some colleges, there is a limitation on the number of new elementary courses that can be introduced at this time.

The Committee believes strongly that mathematics should be taught by mathematicians. Although there are some freshman-sophomore courses in discrete mathematics in computer science departments, the course presented here should, the Committee believes, be taught by mathematicians. The rigor and pace of this course are designed for the freshman level. Some topics necessary for elementary computer science may have to be taught at an appropriate later time, either in a junior-level discrete mathematics course or in the computer science courses.

Needs of Computer Science

What do the computer science majors need? In teaching the first year Introduction to Computer Science course, Tony Ralston kept track of mathematics topics he would have liked the students to have had before (or at least concurrently with) his course:

- *Elementary Mathematics:* Summation notation; subscripts; absolute value, truncation logarithms, trigonometric functions; prime numbers; greatest common divisor; floor and ceiling functions.
- *General Mathematical Ideas:* Functions; sets and operations on sets.
- *Algebra:* Matrix algebra; Polish notation; congruences.
- *Summation and Limits:* Elementary summation calculus; order notation, $0(fn)$; harmonic numbers.
- *Numbers and Number Systems:* Positional notation; nondecimal bases.
- *Logic and Boolean Algebra:* Boolean operators and expressions; basic logic.
- *Probability:* Sample spaces; laws of probability.
- *Combinatorics:* Permutations, combinations, counting; binomial coefficients, binomial theorem.
- *Graph Theory:* Basic concepts; trees.
- *Difference Equations and Recurrence Relations:* Simple differential equations; generating functions.

Many of the ideas are those that students *should have* had in four years of the traditional high school curriculum. In addition, there are some ideas and techniques that are probably beyond the scope of secondary school mathematics. An elaboration of this list appears in the Appendix and in an article by Ralston in *ACM Communications* [33].

Many proposals have been coming from the computer science community. Recommendations for a freshman-level discrete mathematics course from the Educational Activities Board of IEEE probably are the most demanding. Students enrolled in the course outlined in the appendix are first semester freshman also enrolled in the calculus according to the IEEE recommendations published in December 1983, by the IEEE Computer Society [19].

Accreditation guidelines passed recently by ACM and IEEE also require a discrete mathematics course. The recommendations for the mathematics component of a program that would merit accreditation appear below. The criteria appear in their entirety in an article by Michael Mulder and John Dalphin in the April 1984 *Computer* [28].

> Certain areas of mathematics and science are fundamental for the study of computer science. These areas must be included in all programs. The curriculum must include one-half year equivalent to 15 semester hours of study of mathematics. This material includes discrete mathematics, differential and integral calculus, probability and statistics, and at least one of the following areas: linear algebra, numerical analysis, modern algebra, or differential equations. It is recognized that some of this material may be included in the offerings in computer science ....

Presentation of accreditation guidelines which require one and one-half years of study in computer science, one year in the supporting disciplines, one year of general education requirements, and one-half year of electives induced quick and angry response. The liberal arts colleges and the small colleges unable to offer this number of courses or unwilling to require so many credits in one discipline, have responded in many ways. This Small College Task Force of the ACM issued its own report, approved by the Education Board of the ACM [5]. We emphasize only the mathematics portion of those guidelines.

> Many areas of the computer and information sciences rely heavily on mathematical concepts and techniques. An understanding of the mathematics underlying various computing topics and a capability to implement that mathematics, at least at a basic level, will enable students to grasp more fully and deeply computer concepts as they occur in courses .... It seems entirely reasonable and appropriate, therefore, to recommend a substantial mathematical component in the CSIS curriculum .... To this end, a year of discrete mathematical structures is recommended for the freshman year, prior to a year of calculus.

The Sloan Foundation supported representatives of a few liberal arts schools in their attempt to define a high-quality computer science major in such institutions. Again, we put the emphasis on the mathematics component of the proposed program.

From *Model Curriculum for a Liberal Arts Degree in Computer Science* by Norman E. Gibbs and Allen B. Tucker [12]:

> The discrete mathematics course should play an important role in the computer science curriculum .... We recommend that discrete mathematics be either a prerequisite or corequisite for CS2. This early positioning of discrete mathematics reinforces the fact that computer science is not just programming, and that there is substantial mathematical content throughout the discipline. Moreover, this course should have significant theoretical content and be taught at a level

appropriate for freshman mathematics majors. Proofs will be an essential part of the course.

Alfs Berztiss, a member of the Committee, led a number of mathematics and computer science faculty at a conference at the University of Pittsburgh in 1983 at which an attempt was made to formulate a high-quality program in computer science which would prepare good students for graduate study in the field. Details are available in a Technical Report (83-5) from the University of Pittsburgh Department of Computer Science [8]. Both that program and the new and extensive bachelor's program in computer science at Carnegie-Mellon University depend on an elementary discrete mathematics course.

In addition to the proposals for programs, the computer science community is in the process of revising elementary computer science courses. Though old courses stressed language instruction, a more modern approach stresses structured programming and a true introduction to computer science. The beginning courses CS1 and CS2 are described by the ACM Task Force on CS1 and CS2. We quote from the article by Elliot Koffman, *et al.* [21] about the role of discrete mathematics in the structure of these computer science courses.

> We are in agreement with many other computer scientists that a strong mathematics foundation is an essential component of the computer science curriculum and that discrete mathematics is the appropriate first mathematics course for computer science majors. Although discrete mathematics must be taken prior to CS2, we do not think it is a necessary prerequisite to CS1. ...We would ...expect computer science majors and other students interested in continuing their studies in computer science to take discrete mathematics concurrently with the revised CS1.

If high schools and colleges take the recommendation seriously, the student enrolled in CS1 would be enrolled in a discrete mathematics course concurrently. That mathematics course would be required as a prerequisite for CS2. Of all the recommendations, this is likely to have the largest impact on enrollments in discrete mathematics courses.

Syllabus

What are the common needs of mathematics and computer science students in mathematics? The Committee agrees that all the students need to understand the nature of proof, and the essentials of propositional and predicate calculus. In addition, all need to understand recursion and induction and, related to that, the analysis and verification of algorithms and the algorithmic method. The nature of abstraction should be part of this elementary course. While some of the Committee supported the introduction to algebraic structures in this course, particularly for coding theory and finite automata, others felt that those concepts were best left to higher-level courses in mathematics. The basic principles of discrete probability theory and elementary statistics might be considered to be as important and more accessible to students at this level. Professionals in all disciplines cite the importance of teaching problem solving skills. Graph theory and combinatorics are excellent vehicles. All these students need some calculus.

The Committee recommends the inclusion of as many of the proposed topics as possible with the understanding that taste and the structure of the curriculum in each institution will dictate the depth and extent to which they are taught. The ability of students in a course at this level must be considered in making these choices. While one of the goals of the course is to increase the mathematical maturity of the student, some of the mathematical community who have communicated to the panel about their experiences teaching this course have indicated that there are prerequisite skills in reading and in maturity of thinking that really are needed, perhaps even more than in the calculus.

The Committee recognizes that it might be some time before there is as much agreement on the content of a discrete mathematics sequence as there is now about the calculus sequence. In the meantime, diversity and variety should be encouraged so that we may learn what works and what does not. In any case, the Committee strongly endorses the notion that it is not *what* is taught so much as *how*. If the general themes mentioned in the previous paragraph are woven into the content of the course, the course will serve the students well. Adequate time should be allowed for the students to do a lot on their own: they should be solving problems, writing proofs, constructing truth tables, manipulating symbols in Boolean algebra, deciding when, if, and how to use induction, recursion, proofs by contradiction, etc. And their efforts should be corrected.

We have been asked about the role of the computer in this course. To a person we have agreed that this is a mathematics course and that while students might be encouraged, if they have the background, to try the algorithms on a computer, the course should emphasize mathematics. The skills that we are trying to teach will serve the student better than any programming skills we might teach in their place, and the computer science departments prefer it that way. Surely the ideal would be that students be concurrently enrolled in this course and a computer science course where the complimen-

tary nature of the subjects could be made clear by both instructors.

Algorithms are, of course, an integral part of the course. There is still no general agreement on how to express them in informal language. While a form of pseudocode might suit some people, others have found that an informal conversational style suffices. The Committee would not want to make any specific recommendations except that the student be precise and convey his/her methods. It is certainly not necessary to write all algorithms in Pascal. Communication is the key.

The recommendations for a one-year discrete mathematics course are presented in several ways. An outline of the course appears below. In the Appendix, the outline has been expanded to include objectives and sample problems for each topic. The scope and level of the course can be appreciated best from the expanded version.

Discrete Mathematics
A One Year Freshman-Sophomore Course
(*Preliminary Outline*)

Prerequisite: Four years of high school mathematics; may be taken before, during, or after calculus I and II.

1. *Sets*. Finite sets, set notation, set operations, subsets, power sets, sets of ordered pairs, Cartesian products of finite sets, introduction to countably infinite sets.
2. *The Number System*. Natural numbers, integers, rationals, reals, Zn, primes and composites, introduction to operations, and algebra.
3. *The Nature of Proof*. Use of examples to demonstrate direct and indirect proof, converse and contrapositive, introduction to induction, algorithms.
4. *Formal Logic*. Propositional calculus, rules of logic, quantifiers and their properties, algorithms and logic, simplification of expressions.
5. *Functions and Relations*. Properties of order relations, equivalence relations and partitions, functions and properties, into, onto, 1-to-1, inverses, composition, set equivalence, recursion, sequences, induction proofs.
6. *Combinatorics*. Permutations, combinations, binomial and multinomial coefficients, counting sets formed from other sets, pigeon-hole principle, algorithms for generating combinations and permutations, recurrence relations for counting.
7. *Recurrence Relations*. Examples, models, algorithms, proofs, the recurrence paradigm, solution of difference equations.

8. *Graphs and Digraphs.* Definitions, applications, matrix representation of graphs, algorithms for path problems, circuits, connectedness, Hamiltonian and Eulerian graphs, ordering relations—partial and linear ordering, minimal and maximal elements, directed graphs.

9. *Trees.* Binary trees, search problems, minimal spanning trees, graph algorithms.

10. *Algebraic Structures.* Boolean algebra, semigroups, monoids, groups, examples and applications and proofs; or

11. *Discrete Probability and Descriptive Statistics.* Events, assignment of probabilities, calculus of probabilities, conditional probability, tree diagrams, Law of Large Numbers, descriptive statistics, simulation.

12. *Algorithmic Linear Algebra.* Matrix operations, relation to graphs, invertibility, row operations, solution of systems of linear equations using arrays, algebraic structure under operations, linear programming—simplex and graphing techniques.

Preparation for Discrete Mathematics

A consideration of the topics listed in this course outline reveals that, while the course meets our objectives of scope and level, this is a serious mathematics course. The student will have to be prepared for this course by an excellent secondary school background. Those of us who have been teaching freshmen know that many students are coming unprepared for abstract thinking and problem solving. We are aware that many secondary schools are doing a fine job of educating students to handle this work, but many more schools are not. It seems likely that courses ordinarily taught to mathematically deficient first-year students to prepare them for the calculus would also prepare them for this course. In many cases, with only modest changes, these courses can be adapted to be both prediscrete mathematics and precalculus. The Committee expects that the major in computer science will include at least one year of calculus so that at some time the student will surely reap the full benefits of these traditional preparatory courses.

The additional question still remains unanswered—what should be taught in the high schools or on the remedial level in the colleges to prepare students adequately for this course? Our suggestion is tentative: some of us feel that perhaps a revived emphasis on the use of *both* formal and informal proof in geometry courses as a means for teaching methods of proof and analytic thinking would be a step in the right direction.

Others of us are not so sure. Increased use of algorithmic thinking in problem solving could be easily adapted to many high school courses. Readers are encouraged to read Steve Maurer's article in the September 1984 *Mathematics Teacher* for more on this subject.

The Committee on Placement Examinations of the MAA will be attempting to isolate those skills that seem to be needed by students taking discrete mathematics. Although this study might not lead to the development of a placement examination for the course, it will help to explain what might be the appropriate preparation for a successful experience in such a course.

Year after year we face students who claim that they have never seen the binomial theorem, mathematical induction, or logarithms before college. These used to be topics taught at the eleventh or twelfth grade levels. What has happened to them? Students also say that they never had their papers corrected in high school so they never wrote proofs. Some of us have students who cannot tell the hypothesis from the conclusion.

Simple restoration of some of the classical topics and increased emphasis on problem solving might make the proposed course much easier for the student. As one studies the list of topics in the discrete mathematics course, it becomes clear that, in fact, there is little in the way of specific prerequisites for such a course except a solid background in algebra; nothing in the course relies on trigonometry, number theory, or geometry, *per se*. However, the abstraction and the emphasis on some formalism will shock the uninitiated and the mathematically immature.

Recent experimentation at the Sloan-funded schools might tell us something about what we ought to require of students enrolling in this type of course. Results from these schools have not been completely analyzed, but the failure rates seem consistent with those in the calculus courses. Some of the experimental group had taken calculus first and others had not. There seemed to be a filtering process in both cases so that results are not comparable from one discrete mathematics course to another. One Sloan-funded correspondent reported that reading skills might be a factor in success and was following through with a study to see if verbal SAT scores were any indicator of success.

One of the concerns of the Committee throughout its deliberations has been the articulation problem with a course of this kind. We want to be clear that finite mathematics courses in their present form are not the equivalent of this course. We have not totally succeeded in communicating this in presentations at professional meetings. The discrete and finite mathematics courses differ in several ways. First, the discrete mathematics course is not an all-purpose service course. It has been designed primarily for majors in mathematically-related fields. It presumes at least four years of solid secondary school mathematics and hence the level of the course is greater than or equal to the level of calculus. There is inherent in this proposal a heavy emphasis on the use of notation and symbolism to raise the students' ability to cope with abstraction. Secondly, a heavy emphasis on algorithmic thinking is also recommended.

The pace, the rigor, the language, and the level are intended to differ from a standard finite mathematics course. We do not claim that this course can be taught to everyone. Perhaps at some schools the computer science majors are not very high caliber and college programs naturally are geared to the needs of the students. There is nothing inherently wrong in requiring that such students take the mathematics courses required of the business majors: finite mathematics, basic statistics, and "soft" calculus. Perhaps the finite mathematics courses can be improved and sections for some students be enhanced by teaching binary arithmetic and elementary graphs. This is an alternative that many schools will probably choose. It may reflect the reality on a campus where there is really no major in computer science, but a major in data processing or information science which serves its students well. We have not attempted to define that kind of discrete mathematics course. We specifically are defining a course on the intellectual level of calculus for science and mathematics majors. Our visits around the country indicate that many schools need a course at the level of the present finite mathematics offerings. Such courses are a valuable service to some students, but should not be considered equivalent to the course we have described.

Two Year Colleges and High Schools

The mathematics faculty at two year colleges have been working through their own organizations and committees toward curricular reform. The Committee on Discrete Mathematics has attempted to consider their proposals in its own. Jerry Goldstein, Chairperson of CUPM and an ex-officio member of the Committee on the Curriculum at the Two Year Colleges, has been working to maintain articulation between the two groups. The Two Year College Committee began its deliberations after our Committee, so this report reflects only preliminary conclusions from that source. A "Williams"-like conference for the two year colleges took place in the summer of 1984 and proceedings are available from Springer-Verlag in *New Directions in Two Year College Mathematics*, edited by Donald Albers, *et*

al. The situation at this time in the two year colleges is one of exploration, learning, and waiting.

Just as the calculus sequence at two year colleges is taught from the same texts and in the same manner as at the baccalaureate institutions, discrete mathematics courses at two year schools are expected to conform to requirements of four-year schools to which students hoped to transfer. Faculty at Florida State University, in connection with one of the Sloan projects, introduced the discrete mathematics course at a nearby two year college. The course was taught from the same text and in the same manner at both institutions. The students did well and project directors claim the results were "unremarkable."

Recent conferences of the American Mathematical Association of Two Year Colleges (AMATYC) and associations of two year college mathematics faculty in many state organizations have been devoted to the special problems of the two year schools with regard to discrete mathematics. The primary concern of most schools is that they must wait for the four-year schools to indicate what type of course will be transferable. The Committee urges those teaching at four-year institutions to make a special effort to communicate their own requirements to the two year colleges that feed them.

What about discrete mathematics in the high schools? Perhaps it will be an exciting change to see the secondary schools place less emphasis on calculus and more on some of the topics in the discrete mathematics. We understand that there is considerable pressure from parents to have Junior (or Sis) take calculus in high school. We are confident that that will change as the first year of mathematics in the colleges becomes more flexible to include either calculus or discrete mathematics at the same level. If the high schools continue the trend to teaching more computer science for advanced placement, then they will have to offer the discrete mathematics to their students. The present Advanced Placement Examination in computer science is essentially for placement in CS2. To place above CS2, there will probably be a level II examination which conforms to the course outline for CS1 and CS2 as noted in the Koffman report. An Advanced Placement Examination in discrete mathematics is some time in the future, as there is no universal agreement as to exactly what might be included at this time.

In January 1986 a Sloan-funded conference on calculus was held at Tulane University in New Orleans. More than twenty participants presented papers and participated in workshops on the state of calculus and its future. The Committee concurs with that group's consensus that the goals of teaching (mathematics) are to develop increased conceptual and procedural skills, to develop the ability of students to read, write, and explain mathematics, and to help students deal with abstract ideas. These are the global concerns for all mathematics teaching. Secondary schools should be working toward such goals too.

The Committee encourages faculty to get students to work together to solve problems. From experience, some of us have found that students cannot read a problem—either they leave out essential words or do not know how to read the notation when asked to read aloud. The word "it" should be banned from their vocabulary for a while. Students who use the word frequently do so because they do not know what "it" really is. Correcting students' homework has always been one of the best ways of understanding their misconceptions. In discrete mathematics courses this is even more so—concept and procedure vary from problem to problem. Students have to think and be creative. That's tough. They need the re-enforcement of the teacher's comments and the chance to try again. Working with other students should be encouraged because this forces students to speak. This oral communication helps them to learn the terminology and helps them to present clear explanations.

The Impact on Calculus

The concerns of some people that the introduction of discrete mathematics will cause a major change in the calculus will probably prove to be unfounded. However, the Committee believes that there are several important questions to be addressed. We should be asking ourselves if we are doing the best job of teaching calculus. Some of our colleagues outside of mathematics who teach our calculus students have commented to the committee members that there are many aspects of the calculus which seem to be ignored in the present courses. There is widespread dissatisfaction with the problem-solving skills of calculus students. Problems that look even a little different from the ones that they have solved in the standard course are often impossible for students. In addition, we are being held responsible for our students lack of knowledge of numerical techniques. The discrete aspect of the calculus was continually stressed by our respondents. In fact, many commented that we were promoting the idea of a dichotomy in mathematics where there is none by not proposing an integrated program of discrete and continuous mathematics for the first two years. The Committee admits that at this time it is

presenting a feasible solution as opposed to the ideal solution.

Should the teaching of calculus reflect the tremendously powerful symbolic manipulators now on the market? While the most powerful require mainframes, some are available on minicomputers and muMath runs on a personal computer. Can the time previously spent in tedious practice of differentiation or integration be better used to teach the power of the calculus through problem solving and modeling? Two of the Sloan-funded schools—Colby College and St. Olaf College—did experiment with the use of MAC-SYMA, MAPLE, and SMP in the teaching of calculus. At Colby, in a course offered to those who had high school calculus, the computer packages were used to augment the one year single-variable and multivariate calculus course. At St. Olaf, SMP was used in an elective course during a January Interim between the first and second semesters of the standard calculus course. Kathleen Heid writes in *The Computing Teacher* [17] about her experience at the University of Maryland where she taught a section of the "soft" calculus using muMath. The results of all these experiments are quite favorable and indicate an important new consideration in our teaching of the subject.

What about the use of the methods introduced in discrete mathematics in the other courses in the curriculum, including calculus and analysis? What of difference equations? The Committee requested that physical scientists and engineers respond to the idea of changing the calculus. We mentioned the possibility that calculus might contain ideas from discrete mathematics in the solving of traditional calculus-type problems. Several engineers and physicists have responded to our query with some interesting endorsements for change. Those who responded felt that the present mathematical training we offer their professions is inconsistent with what many of them were doing in their jobs—for they were using difference equations and other discrete methods in their everyday applications.

We also should be asking what calculus the computer science major needs. Does the computer science student need the calculus to do statistics and probability? If so, how much rigor is needed? What background is needed in numerical methods? Should mathematics departments be teaching numerical methods? Are the requirements different from numerical analysis? Should we emphasize rigor, technique, or problem-solving skills? Do the traditional courses suffice to encourage integration of discrete and continuous mathematics?

Conclusion

This report is both incomplete and already out-of-date. Questions will continue to arise; answers are not easily found. Textbooks are now being published that are marketed as suitable for elementary discrete mathematics courses. Our annotated bibliography is undoubtedly incomplete. We know of several forthcoming texts that are in manuscript form but which are unlisted because they could not be properly reviewed.

There has been a great deal of interest, much of it enthusiastic, in the revitalization of the elementary college-level mathematics curriculum. The committee members have had the opportunity to visit schools, speak at sectional and national meetings, and to speak personally with hundreds of our colleagues. We are wrestling with problems of ever-changing demands from other disciplines—some, as computer science, so young there is no standard curriculum. We need to adjust our ideals to the realities of our own academic situation. The Committee attempted to propose a course with enough flexibility to allow institutions with different needs to follow the general course outline, putting emphases where they wanted.

The two year colleges and the high schools are dealing with demands of the four-year institutions, parents, and the College Entrance Examination Board. They feel many pressures to keep calculus as the pivotal course. On the other hand, the proposal to integrate discrete mathematics into the high school and even elementary school curricula got considerable support at the 1985 National Council of Teachers of Mathematics (NCTM) meetings in San Antonio.

The recent publication of many discrete mathematics textbooks suitable for the freshman-sophomore year has been exciting. We have the opportunity to see what is successful. The Committee agrees that the next step in the development of the curriculum should be the integration of the discrete and the continuous ideas of mathematics into all courses. That would be ideal and we encourage experimentation to that end.

Committee Members

MARTHA J. SIEGEL, CHAIR, Department of Mathematics, Towson State University; Member of CTUM.

ALFS BERZTISS, Department of Computer Science, University of Pittsburgh; Representative of the ACM Education Board.

DONALD BUSHAW, Department of Pure and Applied Mathematics, Washington State University; Member of CTUM and of CUPM, Chair of MAA Committee on Service Courses.

JEROME GOLDSTEIN, Department of Mathematics, Tulane University; Chair of CUPM.

GERALD ISAACS, Department of Computer Science, Carroll College; Representative of the ACM Education Board.

STEPHEN MAURER, Department of Mathematics, Swarthmore College; then on leave at The Alfred P. Sloan Foundation.

ANTHONY RALSTON, Department of Computer Science, State University of New York at Buffalo; Member of MAA Board of Governors, and organizer of the Williams Conference.

JOHN SCHMEELK, Department of Mathematics, Virginia Commonwealth University; Member of The American Society for Engineering Education (ASEE) Mathematics Education Committee.

Course Objectives and Sample Problems

1. Sets

STUDENT OBJECTIVES:

- Understand set notation.
- Recognize finite and infinite sets.
- Be able to understand and manipulate relations between sets, and make proper use of such terms as:
  - subsets
  - proper subsets
  - supersets
  - equality
  - universe and empty set.
- Understand and be able to manipulate indexed collections of sets.
- Understand and use the set-builder notation.
- Understand and manipulate operations on sets:
  - intersection (finite and countable collections)
  - union (finite and countable collections)
  - difference
  - symmetric difference
  - complement
  - Venn diagrams.
- Understand the proofs of theorems and know the laws:
  - commutative laws
  - associative laws
  - distributive laws
  - DeMorgan's laws.
- Understand Cartesian products of sets and power sets.
- Understand inductive (recursive) definitions of sets.
- Understand a few applications: for example, grammars as sets.
- Be able to do simple proofs by using Venn diagrams or elementary elementwise proofs.

SAMPLE PROBLEMS:

1. List the ordered pairs in the sets

$$A = \{(m, n) \in S \times T : m < n\}$$
$$B = \{(m, n) \in S \times T : m + 1 = n\}$$

where $S = \{1, 2, 3, 4\}$ and $T = \{0, 2, 4, 5\}$.

2. True or false?

$$A \setminus (B \cup C) = (A \setminus B) \cup (A \setminus B).$$

Verify your answer (use elementwise argument, Venn diagram and algebraic manipulation).

3. Let $A_n = \{k \in \mathbf{P} : k \leq n\}$ for each $n \in \mathbf{P}$.

Find $\bigcap_{n=1}^{5} A_n$, $\bigcap_{n=1}^{\infty} A_n$, $\bigcup_{n=1}^{5} A_n$, $\bigcup_{n=1}^{\infty} A_n$.

Find A_n^c and $A_n \cap A_m$ for $n, m \in \mathbf{P}$.

4. For each $n \in \mathbf{N}$, let

$$A_n = \{x \in \mathbf{Q} : x = m/3^n \text{ for some } m \in \mathbf{Z}\}.$$

Describe the set $A_0, A_1, A_2,$ and $A_1 \setminus A_0$.

Find $\bigcap_{n=0}^{\infty} A_n$.

Find $\bigcap_{n=2}^{\infty} A_n$.

5. Sketch the following set S in $\mathbf{N} \times \mathbf{N}$:

$$(0, 0) \in S \text{ and if } (m, n) \in S \text{ then } (m, n + 1) \in S,$$
$$(m + 1, n + 1) \in S \text{ and } (m + 2, n + 1) \in S.$$

Show $S = \{(m, n) : m \leq 2n\}$.

6. Show that if $A \subseteq B$ and $C \subseteq D$, then $A \times C \subseteq B \times D$.

7. Assume A, B, and C are subsets of a universal set U. Simplify

$$(A \cap (B \setminus C))^c \cup A.$$

2. The Number System

STUDENT OBJECTIVES:

- Be able to define
  - positive integers ($\mathbf{P}$)
  - natural numbers ($\mathbf{N}$)
  - integers ($\mathbf{Z}$)
  - rational numbers ($\mathbf{Q}$)
  - irrational numbers
  - reals (as $(-\infty, \infty)$).
- Be able to recognize subsets of $\mathbf{N}$, $\mathbf{P}$, $\mathbf{Q}$, and $\mathbf{Z}$.
- Be able to use interval notation.
- Understand the division algorithm and divisibility.
- Be able to do simple proofs about even and odd numbers (e.g., the sum of two even integers is even).
- Know the definition of prime number, gcd and lcm.
- Be able to find the prime factorization of a number.
- Know how to write an integer given in base 10 as a numeral in base 2.

SAMPLE PROBLEMS:

1. Find elements in the sets (if the set is infinite, list five elements of the set).

$$\{\,n \in \mathbf{N} : n^2 = 4\,\}$$
$$\{\,n \in \mathbf{P} : n \text{ is prime and } 1 \leq n \leq 20\,\}$$
$$\{\,x \in \mathbf{R} : x^2 = 4\,\}$$
$$\{\,n \in \mathbf{P} : n^2 = 3\,\}$$
$$\{\,x \in \mathbf{R} : x^2 \leq 4\,\}$$
$$\{\,x \in \mathbf{R} : x^2 < 0\,\}$$
$$\{\,x \in \mathbf{Q} : x^2 = 3\,\}$$
$$\{\,x \in \mathbf{Q} : 2 < x < 3\,\}$$

2. Determine how many elements are in each set? Write ∞ if the answer is infinite.

$$\{-1, 1\}, \{-1, 0\}, [-1, 1], [-1, 0], P(\mathbf{Z}),$$
$$P([-1, 1]), P(\{-1, 1\}), \{n \in \mathbf{Z} : -1 \leq n \leq 1\},$$
$$\{n \in \mathbf{Z} : -1 < n < 1\}.$$

3. List elements in

$$A = \{\,n \in \mathbf{Z} : n \text{ is divisible by 2}\,\}$$
$$B = \{\,n \in \mathbf{Z} : n \text{ is divisible by 7}\,\}$$
$$C = A \cup B$$
$$D = A \cap B$$

4. Prove that the product of even integers is an even integer.

5. Use the Euclidean Algorithm to determine the greatest common divisor of 741 and 715.

6. Find the prime factorization of 4,978.

7. Determine the numeral in base 2 to represent 81 (base 10).

3. The Nature of Proof

STUDENT OBJECTIVES:

- Be able to identify the hypothesis and the conclusion in sentences of various English constructions.
- Understand the definition of a proposition, its converse, its contrapositive.
- Understand the use of examples as an aid to finding a proof and the misuse of examples as proof.
- Understand the use of counterexamples.
- Be able to do direct proofs, including proof by cases.
- Be able to do indirect proofs.
- Understand the role of axioms and definitions.
- Understand the backward-forward method of constructing a proof.

- Understand and be able to use the principle of mathematical induction.
- See the necessity for the verification of algorithms.
- Do a substantial number of elementary proofs using simple examples from arithmetic.

SAMPLE PROBLEMS:

1. According to the Associated Press, a prominent public official recently said: "If a person is innocent of a crime, then he is not a suspect." What is the contrapositive of this quotation?

2. Prove or disprove the following statement about real numbers x:

$$\text{If } x^2 = x, \text{ then } x = 1.$$

What is the converse of this statement? Prove it or disprove it.

3. After considering some examples if necessary, guess a formula that gives the sum of the interior angles at the vertices of a convex polygon in terms of the number n of sides. Then prove the formula, if you can, by mathematical induction.

4. Write an algorithm for finding the least common denominator of two fractions. Can you think of another?

5. Write an algorithm for finding the median of a list consisting of n (an odd number) real numbers.

6. Prove: if $A \cup B \subseteq A \cap B$, then $A = B$.

7. Prove or disprove: if $A \cap B = A \cap C$, then $B = C$.

8. Prove (by cases): for every $n \in \mathbf{N}$, $n^3 + n$ is even.

9. Prove: for every $n \in \mathbf{P}$,

$$1 + 3 + 5 + \cdots + (2n - 1) = n^2.$$

10. Prove (by contradiction): if x^2 is odd, then x is odd.

11. Prove: There are no integers a and b such that $a^2 = 3b^2$.

12. Prove: $n^3 - n$ is divisible by 3 for every $n \in \mathbf{P}$.

4. Formal Logic

STUDENT OBJECTIVES:

- Write English sentences for logical expressions and vice versa.
- Complete the truth tables for the standard logical connectives.
- Give the truth values of simple propositions given in plain English.
- State the definitions of tautology and contradiction.
- Prove and use the standard logical equivalences: commutative, associative, distributive, and idempotent properties; double negation; DeMorgan laws.

- Recognize computer language commands for standard logical operations.
- State and use logical implications, at least: modus ponens, modus tolens, transitivity of $\rightarrow$ and $\leftrightarrow$.
- Negate $p \vee q$, $p \rightarrow q$, $p \wedge q$.
- Identify the basic quantifiers, free and bound variables, negations and the generalized DeMorgan laws for quantified statements (e.g., $\neg \forall x p(x) \iff \exists x \neg p(x)$).
- Build logic circuits with AND, OR, NOT gates.
- Understand the terms consistency, inconsistency, completeness and decidability (optional).

SAMPLE PROBLEMS:

1. Use a truth table to prove that

   $(p \wedge q) \rightarrow r$ is logically equivalent to $p \rightarrow (q \rightarrow r)$.

2. Prove that $\neg p \wedge r$ is logically equivalent to $\neg(p \vee \neg r)$ without using truth tables.

3. If $p =$ "cows bark", $q =$ "the Orioles are Baltimore's baseball team" and $r =$ "$2 + 4 = 7$", find truth values of $p \wedge q$, $p \rightarrow q$, $(p \wedge q) \rightarrow r$.

4. Consider the proposition for $x \in \mathbb{R}$:

   If $(x - 3)(x - 2) = 0$ then either $x = 3$ or $x = 2$.

   a) Write its converse.
   b) Write its contrapositive.
   c) Write its negation.
   d) What is the truth value of the proposition, its converse, its contrapositive, its negation?

5. Find the result of

   [(0 AND 1) NAND 0] OR NOT [1 IMP 0].

6. Draw a logic circuit representing $(\neg p \wedge q) \vee r$.

7. Prove the following logical argument: $p \wedge q$, $p \rightarrow r$, $\neg s \rightarrow q$, and $s \rightarrow t$ imply $r \wedge t$.

8. Determine truth values of the following propositions. Assume the universe is $\mathbb{N}$.
   (a) $\forall m \exists n[m = n^2]$ (b) $\exists m \forall n[m = n^2]$

9. Write in logical form: for every $x, y \in \mathbb{R}$, there exists $z \in \mathbb{R}$ such that $x < z$ and $z < y$.

10. Negate $\forall x[p(x) \wedge \neg q(x)]$.

11. Answer each of the following in the appropriate box.
    (a) If the book costs more than \$20, it is a best-seller. The book costs more than \$20. Is the book a best-seller? ☐ yes, ☐ no, ☐ not enough information.
    (b) If the kite is multicolored, it will fly. The kite flies. Is the kite multicolored? ☐ yes, ☐ no, ☐ not enough information.

(c) If the bed is comfortable, Sally will sleep. The bed is not comfortable. Will Sally sleep? ☐ yes, ☐ no, ☐ not enough information.

(d) If the candidate is elected in Vermont, she will be elected by the country. The candidate is not elected by the country. Is the candidate elected in Vermont? ☐ yes, ☐ no, ☐ not enough information.

12. Simplify the logic circuit below:

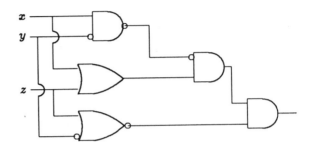

5. Functions and Relations

STUDENT OBJECTIVES:

- Be able to define "function" and "relation".
- Know the properties of relations:
  reflexive
  transitive
  symmetric
  antisymmetric.
- Be able to identify order relations.
- Be able to identify equivalence relations.
- Understand the relationship between equivalence relations and partitions.
- Know the definitions of domain, codomain, image, into, onto (or surjection), one-to-one (or injection), bijection.
- Be able to do simple proofs involving these definitions.
- Be able to work with composition and inverses of relations and, in particular, of functions.
- Be familiar with recursive definitions of functions.
- Be introduced to sequences as functions, again with some emphasis on recurrence relations and recursion.
- Be able to do proofs involving recursion.
- Be able to work with definitions of relations as ordered pairs as opposed to as "rules".
- Know the definition of the characteristic function of a set.

SAMPLE PROBLEMS:

1. Give at least one reason why each of the following does *not* define an equivalence relation on the set of integers:
   a) $x + y$ is odd;
   b) $x < 2y$.

2. Recall that a positive integer is *prime* if it has exactly two positive integer divisors: itself and 1. Consider the relation defined on the set of all integers greater than 1 by: "y is the smallest prime that is a divisor of x."
   a) Explain why this relation is a function $y = f(x)$.
   b) What is the range of f?
   c) List four elements of $f^{-1}(5)$.
   d) Prove that $f \circ f = f$.

3. When the prevailing rate of interest is $100r\%$, an account that has P dollars in it at the beginning of a year should have how much in it at the beginning of the next year? Express your answer as a recursion formula, and solve it to find the size of such an account after n years.

4. The *factorial*, usually denoted by $n!$, of a positive integer n is the product of all positive integers from 1 to n inclusive. Show how $n!$ may be defined recursively.

5. *Prove* that $f(x) = 2x + 1$ is one-to-one and onto from **R** to **R**. Is f one-to-one from **Z** to **Z**? Does f map **Z** onto **Z**? Verify your answers.

6. List five elements in the sequence given by $a_0 = 1$, and $a_n = 2a_{n-1}$ for $n \geq 1$). Give another formula for a_n, for any $n \in$ **N**.

7. Let $\Sigma = \{a, b\}$ and let Σ^* be the set of words over Σ. If w_1 and w_2 are elements of Σ^*, define $w_1 \leq w_2$ if and only if length $(w_1) \leq$ length (w_2). Is $\leq$ a partial order? Why?

8. Prove: If $h(1) = 1$ and $h(n + 1) = 2 \cdot h(n) + 1$ for $n \geq 1$, then $h(n) = 2^n - 1$ for all $n \in$ **P**.

9. For $m, n \in$ **N** define

$m \sim n$ if and only if $m^2 - n^2$ is divisible by 3.

Prove that $\sim$ is an equivalence relation on **N**. Find 8 elements of each of the equivalence classes $[0]$ and $[1]$. What is the partition of **N** induced by $\sim$?

6. Combinatorics

STUDENT OBJECTIVES:

- Be able to apply the basic permutation and combination formulas.
- Be familiar with and be able to provide basic combinatorial identities using combinatorial reasoning.

- Be able to use the binomial theorem.
- Be able to do ball and urn type problems.
- Be able to state and apply the inclusion-exclusion principle.
- Be able to apply the pigeon-hole principle.
- Be familiar with combinatorial algorithms based on recurrence relations.
- Be introduced to the basic ideas of intuitive discrete probability.

SAMPLE PROBLEMS:

1. In many states automobile license plates consist of three (capital) letters followed by three digits. Are there any states in which this probably does not give enough different license plates even if discarded plate numbers can be reused? Are there any states in which three letters followed by three digits or three digits followed by three letters is probably not enough? How many license plates are possible in your state?

2. In a hypnosis experiment, a psychologist inflicts a sequence of flashing lights on a subject. The psychologist has red, blue and green lights available. How many different ways are there to inflict 9 flashes if two are red, four blue and three green?

3. How many triangles are there using edges and diagonals of an n-sided polygon if the vertices of the triangle must be vertices of the polygon?

4. Verify by induction and by a combinatorial argument that

$$\sum_{k=m}^{n} C(k, m) = C(n + 1, m + 1).$$

What does this say about Pascal's triangle?

5. Evaluate

$$\sum_{k=0}^{n} k^2 C(n, k).$$

6. How many ways are there to take four distinguishable balls and put two in one distinguishable urn and 2 in another if
   a) the order in which the balls are put in the urns makes a difference;
   b) the order does not make a difference.

7. How many integers less than 105 are relatively prime to 105?

8. Use the algorithm which generates all permutations of length n of $1, 2, \ldots, n$ where no digit can be repeated to derive an algorithm to generate all permutations when any digit may be repeated an arbitrary number of times.

7. Recurrence Relations

STUDENT OBJECTIVES:

- Have lots of exercise in the elements of recursive thinking (e.g., the recursive paradigm—solve problems by jumping into the middle and working your way out).
- Be familiar with recursive definitions of syntax.
- Be familiar with recursive algorithms.
- Understand what a difference equation is.
- See how difference equations can be used to model practical problems.
- Understand the methods for the solution of linear, constant coefficient equations and first order difference equations.
- Be familiar with some applications of difference equations.
- Be able to use the difference calculus (optional).

SAMPLE PROBLEMS:

1. Compute the first ten terms of the sequence defined by

$$\text{if } n = 1 \text{ or } 2 \text{ then } f_n = 3$$
$$\text{else } f_n = f_{n-1}^2 + f_{n-2}.$$

2. What is $f(1)$ if f is defined by

$$f(n) = \begin{cases} n - 3 & \text{if } n \geq 1000 \\ f(f(n + 6)) & \text{if } n < 1000. \end{cases}$$

3. Describe the strings of characters defined by

$$\langle word \rangle ::= \langle digit \rangle | \langle letter \rangle \langle word \rangle \langle letter \rangle$$

with the standard definitions of *digit* and *letter* and where := is read "is defined to be".

4. Consider the sum

$$1 + 3 + 5 + 7 + \ldots + 2n - 1.$$

Use the inductive and recursive paradigms to conjecture a closed form expression for this sum.

5. Suppose we add to the usual form of the Towers of Hanoi problem the rule that a disk can only be moved from one peg to an adjacent peg (i.e., you can never move a disk from peg 1 to peg 3 or vice versa). Devise an algorithm for solving this version of the problem. Display solutions when $n = 2, 3, 4$.

6. Display a recursive version of the Euclidean algorithm.

7. Consider the recursion

$$P_n = 1 + \sum_{k=1}^{n-1} P_k \qquad n > 1, \qquad P_1 = 1.$$

Compute several terms. Find the pattern and prove that it is correct.

8. Consumer loans work as follows. The Lender gives the Consumer a certain amount P, called the Principal. At the end of each payment period (usually each month) the consumer pays the Lender a fixed amount p. This continues for a prearranged number of periods (e.g., 60 months = 5 years). The value of p is calculated so that, at the end of the time, the Principal and all interest due have been paid off exactly. During each payment period the amount owed by the Consumer increases by r, the period interest rate, but it also decreases at the end of the period by p. Let P_n be the amount owed after the nth payment is made. Find a difference equation and boundary conditions for P_n.

9. Solve

$$a_{n+1} = 5a_n - 6a_{n-1}, a_1 = 5, a_2 = 7.$$

10. Find the general solution to

$$a_n - 3a_{n-1} + 2a_{n-2} = n2^n.$$

11. Solve

$$S_n = (n-1)/nS_{n-1} - 1/[n^2(n-1)].$$

12. In ternary search, t and u are the entries closest to 1/3 and 2/3 of the way through the list. Let the search word be w. If $w < t$, then search the first third of the list by ternary search. Similarly, if $t < w < u$ or $u < w$ search, respectively, the middle and last thirds of the list. Write recurrence relations for the worst and average case number of comparisons in ternary search. Also show that, if an appropriate sequence $\{n_i\}$ of list lengths is chosen, then W_n, the number of comparisons in the worst case is given by

$$W_n = 2 \log_3(n + 1).$$

(This problem assumes that students have seen a similar analysis for binary search.)

8. Graphs and Digraphs

STUDENT OBJECTIVES:

- Understand the definition of the digraph and its use as the picture of a relation.
- Be able to write the matrix representation of a digraph.
- See many applications of digraphs as natural models for networks in real life, such as systems of roads, pipelines, airline routes.
- Know the definitions: connectedness, completeness, complement.
- Be introduced to path problems (and transitive closure) and Warshall's algorithm.
- Be familiar with undirected graphs, the associated definitions and the classical problems of graph theory—the bridges of Koenigsberg, the four color problem, Kuratowski's theorem.
- Be encouraged to solve interesting problems like *Mastermind* and *Instant Insanity* to see the usefulness of graph theory.
- Be using algorithms such as Kruskal's algorithm and Dijkstra's algorithm in solving problems.
- Have the opportunity to see the applications to activity analysis (CPM and PERT).
- Be exposed to depth-first search algorithms and topological sorting.

SAMPLE PROBLEMS:

1. Prove that a connected graph of n nodes contains at least $n - 1$ edges.
2. Prove that a digraph is disconnected iff its complement is connected. (The complement of a digraph D is defined by the matrix obtained when in the adjacency matrix of D every 0 is replaced by a 1, and every 1 by a 0.)
3. A digraph $D = \langle A, R \rangle$ is complete if for all $a, b \in A$, $\langle a, b \rangle \in R$ implies $\langle b, a \rangle \in R$. With respect to this definition, is it true that $\langle a, a \rangle \in R$ for all $a \in A$, or is it true that $\langle a, a \rangle \notin R$ for all $a \in A$?
4. A digraph $D = \langle A, R \rangle$ is a tournament if, for all $\langle a, b \rangle \in A$, $\langle a, b \rangle \in R$ or $\langle b, a \rangle \in R$ whenever $a \neq b$, but $\langle a, b \rangle \in R$ implies $\langle b, a \rangle \notin R$. How many tournaments are there as a function of n, where $n = |A|$? Draw all tournaments for $n = 3$.
5. Prove (by induction) that every tournament contains a Hamiltonian path.
6. How many digraphs on n nodes are there? How many graphs?
7. Find the shortest path from node 1 to every other node in a specific given digraph.
8. Find the transitive closure of the relation represented by this same digraph.

9. Trees

STUDENT OBJECTIVES:

- Know the definition of a tree.
- Be able to find the minimal spanning trees for a given graph.
- See the many applications of trees in search problems, with a complete introduction to binary search trees.
- See how to convert digraphs to trees.
- Know how to use digraph algorithms for cycles and critical path analysis.
- Prove the theorems on trees by induction and rely on recursive algorithms for transversal problems on trees.
- Be familiar with sorting and searching algorithms.
- See rooted trees and Polish notation as an application.

SAMPLE PROBLEMS:

1. Show the equivalence of the following definition of an undirected tree: (a) a connected graph without any circuits; (b) a connected graph that becomes disconnected on the removal of any one edge; (c) a connected graph with its number of edges one less than its number of nodes.
2. Prove that the number of leaves in a binary tree with n internal nodes is at most $n + 1$.
3. Prove that for every nonnegative integer n it is possible to construct a binary tree with n leaves in which the outdegree of every internal node is 2.
4. Find all spanning trees of a specific given graph.
5. Find all simple cycles in a specific given graph.
6. Given the drawing of a scheduling network, find the critical path(s) in this network.

10. Algebraic Structures

STUDENT OBJECTIVES:

- Be able to define and recognize unary and binary operations.
- Be able to distinguish whether sets are closed with respect to a given operation.
- Be familiar with a variety of operations on a variety of sets:

    arithmetic operations on $\mathbf{N}, \mathbf{P}, \mathbf{Q}, \mathbf{Z}, \mathbf{R}$
    set operations on $P(S)$
    logical operations on propositions
    matrix operations on 2×2 matrices.

- Be able to understand the general definition of an operation on a set via some unfamiliar rule or a table.
- Be able to decide which of the properties hold:

commutative
associative
existence of identity
existence of inverse for given operations on given sets.

- Recognize semigroups, monoids, groups, and cosets.
- Have an elementary knowledge of finite group codes (need cosets).
- Be familiar with and be able to manipulate boolean algebras with many examples.
- Have a rudimentary knowledge of lattices.
- Be able to apply the ideas of homomorphism and isomorphism.

SAMPLE PROBLEMS:

1. Determine if $(\mathbf{N}, +)$ is a group.
2. Determine if $(Z, \circ)$ is a semigroup, a monoid, a group, when $a \circ b$ is defined to be $a + b - 2$ whenever $a, b \in Z$.
3. For $X = 011010$ and $Y = 100100$, find the Hamming distance from X to Y.
4. A code has a minimum distance of 6. How many errors can it detect? How many errors can it correct?
5. Determine if $(\mathbf{Z}_5, *)$ is a group.
6. Let $S = \{a, b, c\}$ and define the operation $\oplus$ by the table below. Determine whether $(S, \oplus)$ is a semigroup.

$\oplus$	a	b	c
a	a	c	b
b	b	b	a
c	b	a	c

7. Prove that in every boolean algebra, $[B, +, \circ, ', 0, 1]$, $x \circ x = x$ and $x + 1 = 1$ for any $x \in B$.
8. Show that the set of 2×2 matrices with integer entries is a commutative monoid under matrix addition.
8. Construct a logic network for the the boolean expression
$$x_1 \circ x_2' + (x_1 \circ x_3)'.$$

11. Discrete Probability and Descriptive Statistics

STUDENT OBJECTIVES:

- Understand basic axioms, simple theorems of probability.
- Understand conditional probability.
- Understand, and be able to do problems involving the discrete uniform, Bernoulli, binomial, Poisson (optional), hypergeometric and geometric probability distributions and their random variables.
- Understand the goal of random number generation.
- Understand the Law of Large Numbers.

- Have a working knowledge of descriptive statistics:
  populations vs. samples
  simple graphing techniques
  calculation and meaning of mean, median and mode
  calculation and meaning of standard deviation (calculations may be restricted to ungrouped data).
- Discuss the interpretation of sample data including exploratory data analysis.
- Know the meaning of expected value and variance for random variables.
- Know how to use Chebyshev's inequality.
- Know how to simulate some of the probability models discussed.
- Should be doing problems that demonstrate the relationship between probability and difference equations, especially through classical problems like the gambler's ruin problem.

SAMPLE PROBLEMS:

1. Find the probability that in a random arrangement of n files, we find them to be in alphabetical order.
2. Simulate the gambler's ruin problem assuming that the coin being tossed is fair, that A wins a dollar from B when the coin lands heads up and gives a dollar to B otherwise. You may assume that A begins the game with \$3 and B begins with \$2. Determine the average length of a game and determine the frequency with which A wins the game.
3. Solve the gambler's ruin problem using a probability model. Use the situation given in problem two and compare your results here to the results of the simulation above.
4. If scores on an examination are normally distributed with $\mu = 500$ and $\sigma = 75$, find the probability that a score exceeds 700. Find the 95th percentile for the scores. Find the probability that a sample mean of more than 510 is found for a random sample of 100 scores.
5. Given the data below, sketch a stem and leaf display, a frequency histogram, and find the mean, median, mode and sample standard deviation. Determine the 75th percentile of these data.
6. A prize has been put in 2% of all Sweeties cereal boxes. Find the probability that the fourth box you open contains the first prize you find. Determine the average number of boxes one needs to open in order to get one prize. Simulate the experiment, also.
7. If 80% of all programs fail to run on the first try, find the probability that in a group of 100 programs

at least 30 programs run on the first try.

8. Show $b(x; n, 1-p) = b(n-x; n, p)$ where $b(x; n, p) = \binom{n}{x} p^x (1-p)^{n-x}$.

9. Completion time on a standardized test is normally distributed with an average of 40 minutes and standard deviation 5 minutes. How much time should be allotted if the examiner wants 95% of the students to finish the test? What if the examiner wishes to leave 5 minutes for checking for 95% of the class?

12. Algorithmic Linear Algebra

STUDENT OBJECTIVES:

- Understand matrix operations and their properties.

- Be able to determine whether a matrix is invertible, and if so, be able to find the inverse.

- See the relationship of matrices to graphs.

- See the use of matrices in representation of linear systems.

- Be able to use row operations to reduce matrices.

- Be able to determine whether a system of linear equations has a solution, a unique solution or no solution.

- Be able to solve a system of linear equations, if a solution exists.

- Have an understanding of linear inequalities, graphing them in the two variable case.

- Be able to solve linear programming problems using the simplex method (and the graphical method in the two variable case).

- Be able to use matrices to solve Markov chain models.

- Use powers of incidence matrices to study connectivity properties of graphs or digraphs.

- See and use the recursive definition of the determinant of a square matrix.

SAMPLE PROBLEMS:

1. Determine the matrix representation of the undirected graph pictured below. Determine the number of paths of length 3 from v_1 to v_4.

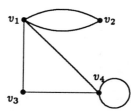

2. A firm packages nut assortments: Fancy & Deluxe. The Fancy assortment contains 6 oz. cashews, 8 oz. almonds and 10 oz. peanuts. It sells for $2.40. The Deluxe assortment contains 12 oz. cashews, 10 oz. almonds and 8 oz. peanuts. It is priced at $3.60. The supplier can provide a maximum of 3000 oz. cashews, 3600 oz. almonds and 3200 oz. peanuts. Find the number of boxes of each type that would maximize revenue. Use the simplex method and a graphical method.

3. Determine if each of the systems

(a) $4x + 3y = 7$
     $2x + 6y = 8$

(b) $6x + 3y + 7z = 4$
     $2x + 5y + 8z = 10$

has one solution, no solution, many solutions. Find all solutions in each case.

4. Let

$$A = \begin{bmatrix} 2 & 1 \\ 3 & 4 \end{bmatrix} \qquad B = \begin{bmatrix} 1 & 0 \\ 0 & 0 \end{bmatrix}$$

$$C = \begin{bmatrix} 1 & 0 & 1 \\ 3 & -1 & 2 \end{bmatrix} \qquad D = \begin{bmatrix} 1 & 0 & 1 \\ 2 & 1 & 1 \\ 0 & 3 & 1 \end{bmatrix}$$

Determine if the following exist. If they do, find them; if not, explain. $A + B$, AB, $A + D$, A^{-1}, B^{-1}, D^{-1}, $\det A$, $\det C$.

5. Given a Markov chain with transition matrix P, find the steady state probability vecotr. Let

$$P = \begin{bmatrix} .3 & .7 \\ .4 & .6 \end{bmatrix}.$$

(Here we would give a word problem with this transition matrix.)

Bibliography

Textbooks

This is a list of textbooks that might be considered for use in courses of the kind discussed in this report. It is as nearly complete as we could make it, but books of this kind are still appearing often and we may well have overlooked some good older ones. Inclusion in the list thus does not imply endorsement, nor does omission imply the opposite. Likewise, the "notes" are not meant to be definitive in any way, but just remarks that users of this bibliography may find interesting.

1. Arbib, M.; Kfoury, A.; Moll, R. *A Basis For Theoretical Computer Science*. New York: Springer-Verlag; 1981; ISBN 0-387-90573-1.

   An introduction to theoretical computer science. Chapter 4 includes techniques of proving theorems. Contents: Sets, maps, and relations; induction, strings, and languages; counting, recurrence, and trees; switching circuits, proofs, and logic; binary relations, lattices, and infinity; graphs, matrices, and machines.

2. Biggs, Norman L. *Discrete Mathematics*. New York: Oxford University Press; 1985; ISBN 0-19-853252-0.

   A mathematically-sound book with plenty of material for a two-semester course. Three main sections are "Numbers and counting," "Graphs and algorithms," and "Algebraic methods." Although written from the viewpoint of mathematics rather than computer science, it does pay a fair amount of attention to algorithms. Probably a bit too rigorous for freshmen and sophomores. Contents: Graphs, combinatorics, number theory, coding theory, combinatorial optimization, abstract algebra.

3. Bogart, Kenneth. *Introductory Combinatorics*. Boston: Pitman; 1983; ISBN 0-273-01923-6.

   A fairly complete coverage of standard combinatorial topics, but the treatment is essentially non-algorithmic; e.g., there is no algorithm for permutations. Somewhat sophisticated; probably requires a year of calculus for maturity. Contents: Introduction to enumeration; equivalence relations, partitions, and multisets; algebraic counting techniques; graph theory; matching and optimization; combinatorial designs; partially ordered sets.

4. Brualdi, Richard A. *Introductory Combinatorics*. New York, etc.: North-Holland; 1977; ISBN 0-7204-8610-6.

   Sophomore level; calculus prerequisite. Sophisticated, but not much algorithmic flavor. Contents: What is combinatorics? the pigeonhole principle; basic counting principles: permutations and combinations; the binomial coefficients; the inclusion-exclusion principle; recurrence relations; generating functions; systems of distinct representatives; combinatorial designs; introduc-

tion to the theory of graphs; chromatic number, connectivity, and other graphical parameters; optimization problems.

5. Cohen, Daniel I.A. *Basic Techniques of Combinatorial Theory*. New York: John Wiley and Sons; 1978; ISBN 0-471-03535-1.

   Assumes one semester of calculus; not inclined to use proof by induction. Contents: Introduction; binomial coefficients; generating functions; advanced counting numbers; two fundamental principles; permutations; graphs. Appendix on mathematical induction.

6. Dierker, Paul; Voxman, William. *Discrete Mathematics*. San Diego: Harcourt Brace Jovanovich; 1986; ISBN 0-15-517691-9.

   College algebra a prerequisite; primarily for freshmen and sophomores. The theme of algorithms is a unifying thread; otherwise, little independence between chapters, so could be used as a text for one- or two-semester courses. Contents: A first look at algorithms; number systems and modular arithmetic; introduction to graph theory; applications of graph theory; boolean algebra and switching systems; symbolic logic and logic circuits; difference equations; an introduction to enumeration; elementary probability theory; generating functions; introduction to automata and formal languages; appendices on set theory, functions, matrices, and relations.

7. Doerr, Alan; Levasseur, Kenneth. *Applied Discrete Structures for Computer Science*. Chicago: Science Research Associates; 1985; ISBN 0-574-21755-X.

   Aimed at freshman-sophomore computer science majors. Includes applications, some "Pascal notes." Contents: Set theory; combinatorics; logic; more on sets; introduction to matrix algebra; relations; functions; recursion and recurrence relations; graph theory; trees; algebraic systems; more matrix algebra; boolean algebra; monoids and automata; group theory and applications; an introduction to rings and fields.

8. Gersting, Judith. *Mathematical Structures for Computer Science*. San Francisco: W.H. Freeman and Company; 1982; ISBN 0-7167-1305-5.

   A fine text, but emphasis on computer science applications may be too great. An accessible reference on group codes. Contents: How to speak mathematics: basic vocabulary; structures and simulations; boolean algebra and computer logic; algebraic structures; coding theory; finite-state machines; machine design and construction; computability; formal languages.

9. Grimaldi, Ralph P. *Discrete and Combinatorial Mathematics*. Reading, Mass.: Addison-Wesley; 1985; ISBN 0-201-12590-0.

   Intended for sophomores and juniors. Contents: Fundamental principles of counting; enumeration in set theory; relations and functions; languages; finite state machines; relations: the second time around; the system of integers; the principle of inclusion and exclusion; rings and modular arithmetic; boolean algebra and switching functions; generating functions; recurrence relations; groups, coding theory, and Pólya's method of enumeration; finite fields and combinatorial designs; an introduction to graph theory; trees; optimization and matching.

10. Hillman, Abraham P.; Alexanderson, Gerald L.; Grassl, Richard M. *Discrete and Combinatorial Mathematics*. New York: Dellen Publishing Company; 1986; ISBN 0-02-354580-1.

    Sophomore-junior text. Contents: Sets and relations; algebraic structures; logic; induction; combinatorial principles; digraphs and graphs; groups; polynomials and rational functions; generating functions and recursions; combinatorial analysis of algorithms; introduction to coding; finite state machines and languages.

11. Johnsonbaugh, R. *Discrete Mathematics, Revised Edition*. New York: Macmillan; 1984; ISBN 0-02-360900-1.

    Intended for a one-semester course for freshmen or sophomores. Mainly but not exclusively aimed at computer science students. Emphasizes an algorithmic approach and does a considerable amount of algorithm analysis. Contents: Introduction; counting methods and recurrence relations; graph theory; trees; network models and Petri nets; boolean algebras and combinatorial circuits; automata, grammars, and languages. Appendices on logic and matrices.

12. Kalmanson, Kenneth. *An Introduction to Discrete Mathematics and Its Applications*. Reading, Mass.: Addison-Wesley; 1986; ISBN 0-201-14947-8.

    Intended for freshmen and sophomores. Developed in conjunction with Sloan-funded course at Montclair State College. Contents: Sets, numbers, and algorithms; sets, logic and computer arithmetic; counting; introduction to graph theory; trees and algorithms; directed graphs and networks; applied modern algebra; further topics in counting and recursion; appendix with programs in BASIC.

13. Kolman, Bernard; Busby, Robert C. *Discrete Mathematical Structures for Computer Science*. Englewood Cliffs, New Jersey: Prentice-Hall; 1984; ISBN 0-13-215418-8.

    "There are no formal prerequisites, but the reader is encouraged to consult the Appendix as needed." Intended for a one- or two-semester course for freshmen or sophomore computer science students. Contains relatively few algorithms. Approach on the informal side, with not very many theorems or proofs. Contents: Fundamentals; relations and digraphs; functions; order relations and structures; trees and languages; semigroups and groups; finite-state machines and languages; groups and coding.

14. Korfhage, Robert R. *Discrete Computational Structures, Second Edition*. New York: Academic Press; 1984; ISBN 0-12-420860-6.

    Contents: Entities, properties, and relations; arrays and matrices; graph theory: fundamentals; combina-

torics; trees and hierarchies; graph theory: undirected graphs; graph theory: directed graphs; discrete probability; automata and formal languages; boolean algebras; logic: propositional and predicate calculus; algorithms and programs.

15. Levy, Leon S. *Discrete Structures of Computer Science*. New York: John Wiley and Sons; 1980; ISBN 0-471-03208-5.

    A highly-personal statement on discrete structures for computer science students. The presentation is very sketchy. For a sophomore-junior course; leaps quickly into abstraction and algorithms. Contents: An essay on discrete structures; sets, functions, and relations; directed graphs; algebraic systems; formal systems; trees; programming applications.

16. Lipschutz, Seymour. *Discrete Mathematics*. New York: McGraw-Hill (Schaum's Outline Series); 1976; ISBN 0-07-037981-5.

    Contains an outstanding collection of (easy) worked examples and exercises. Vectors and matrices are regarded as an introductory topic. Contents: Set theory; relations; functions; vectors and matrices; graph theory; planar graphs, colorations, trees; directed graphs, finite-state machines; combinatorial analysis; algebraic systems, formal languages; posets and lattices; proposition calculus; boolean algebra.

17. Lipschutz, Seymour. *Essential Computer Mathematics*. New York: McGraw-Hill (Schaum's Outline Series); 1982; ISBN 0-07-037990-4.

    Again there is an excellent collection of examples and exercises. Includes discussion of representation of numbers and characters, linear algebra, and probability and statistics. Suitable for technical mathematics course for data processing students; not appropriate as a text for the discrete mathematics course proposed by the Committee. Contents: Binary number system; computer codes; computer arithmetic; logic, truth tables; algorithms, flow charts, pseudocode programs; sets and relations; boolean algebra, logic gates; simplification of logic circuits; vectors, matrices, subscripted variables; linear equations; combinatorial analysis; probability; statistics: random variables; graphs, directed graphs, machines.

18. Liu, C.L. *Elements of Discrete Mathematics, Second Edition*. New York: McGraw-Hill; 1985; ISBN 0-07-038133-X.

    A comparatively short, but well-constructed text. Intended for a one-semester course but probably more appropriate for juniors than for freshmen or sophomores, although there is no prerequisite beyond high school algebra. Induction and problem-solving are treated early. A rather traditional mathematical approach with emphasis on combinatorics, relatively little on algorithms. Contents: Computability and formal languages; permutations, combinations, and discrete probability; relations and functions; graphs and planar graphs; trees and cutsets; finite-state machines; analysis of algorithms; discrete numeric functions and generating functions; recurrence relations and recursive algorithms; groups and rings; boolean algebras.

19. Liu, C.L. *Introduction to Applied Combinatorial Mathematics*. New York: McGraw-Hill; 1968; ISBN 0-07-038124-0.

This is a good source for recurrence relations, and for Pólya's theory of counting. It also contains introductions to linear and dynamic programming. Contents: Permutations and combinations; generating functions; recurrence relations; the principle of inclusion and exclusion; Pólya's theory of counting; fundamental concepts in the theory of graphs; trees, circuits, and cut-sets; planar and dual graphs; domination, independence, and chromatic numbers; transport networks; matching theory; linear programming; dynamic programming; block designs.

20. Marcus, Marvin. *Discrete Mathematics: A Computational Approach Using BASIC*. Rockville, Maryland: Computer Science Press; 1983; ISBN 0-914894-38-2.

Interesting approach; elementary. Complemented by a DOS 3.3 16-sector 5 1/4" floppy disk, DISCRETE PROGRAMS. Contents: Elementary logic; sets; relations and functions; some important functions; function optimization; induction and combinatorics; introduction to probability; introduction to matrices; solving linear equations; elementary linear programming.

21. Molluzzo, John L.; Buckley, Fred. *A First Course in Discrete Mathematics*. Belmont, Calif.: Wadsworth Publishing Company; 1986; ISBN 0-534-05310-6.

"...intended for non-mathematically-oriented students ...first or second-year computer science or computer information systems student." Contents: Number systems; sets and logic; combinatorics; probability; relations and functions; vectors and matrices; boolean algebra; graph theory; appendix on Pascal.

22. Mott, Joe L.; Kandel, Abraham; Baker, Theodore P. *Discrete Mathematics for Computer Scientists*. Reston, Virginia: Reston Publishing Company; 1983; ISBN 0-8359-1372-4.

Sophomore-junior course; programming experience desirable but not essential. For computer science audience (posets are defined on p. 17). Contents: Foundations; elementary combinatorics; recurrence relations; relations and digraphs; graphs; boolean algebras.

23. Norris, Fletcher R. *Discrete Structures: An Introduction to Mathematics for Computer Scientists*. Englewood Cliffs, New Jersey: Prentice-Hall; 1985; ISBN 0-13-215260-6 (Instructor's Manual; ISBN 215277).

Written for a one-semester course for freshmen and sophomores. College algebra is the prerequisite. Contents: Propositions and logic; sets; boolean algebra; the algebra of switching circuits; functions, recursion, and induction; relations and their graphs; applications of graph theory; discrete counting: an introduction to combinatorics; posets and lattices; appendices on the binary number system and matrices.

24. Pfleeger, Shari Lawrence; Straight, David W. *Introduction to Discrete Structures, Revised Edition*. New York: John Wiley and Sons; 1985; ISBN 0-471-80075-9.

Aimed at computer science majors; no college-level prerequisites; theory with applications. Includes some proofs. Contents: Formal systems; functions and relations; boolean algebras; boolean algebra and logic design; lattices and their applications; cardinality and countability; graphs and their use in computing; introduction to formal languages; computability.

25. Polimeni, Albert D.; Straight, H. Joseph. *Foundations of Discrete Mathematics*. Monterey, Calif.: Brooks Cole Publishing Company; 1985; ISBN 0-534-03612-0.

Intended for sophomores. Prerequisite: one year of college-level mathematics, including a semester of calculus, and an introductory programming course. Pascal used throughout. Contents: Logic; set theory; number theory and mathematical induction; relations; functions; algebraic structures; graph theory.

26. Prather, Ronald P. *Discrete Mathematical Structures for Computer Science*. Boston: Houghton Mifflin; 1976; ISBN 0-395-20622-7 (Solutions Manual; ISBN 0-395-20623-5).

A solid coverage of all the standard material. Boolean algebras are treated as lattices. Contents: Preliminaries, algebras and algorithms, graphs and digraphs, monoids and machines, lattices and boolean algebras, groups and combinatorics, logic and languages.

27. Prather, Ronald P. *Elements of Discrete Mathematics*. Boston: Houghton Mifflin; 1986; ISBN 0-395-35165-0 (Solutions Manual; ISBN 0-395-35166-9).

Suitable for a one-term course. "No prior programming experience is needed because a generic pseudocode language is used to phrase algorithms." Developed under a Sloan Foundation pilot project grant. Contents: Intuitive set theory; deductive mathematical logic; discrete number systems; the notion of an algorithm; polynomial algebra; graphs and combinatorics.

28. Roman, Steven. *An Introduction to Discrete Mathematics*. Philadelphia: Saunders College Publishing; 1986; ISBN 0-03-064019-9.

Could be used for a one- or two-semester course for freshmen or sophomores in mathematics as well as computer science. A careful and not too hurried approach but quite traditionally mathematical with little attention to algorithms. Contents: Sets, functions, and proof techniques; logic and logic circuits; relations on sets; combinatorics—the art of counting; more on combinatorics; an introduction to graph theory.

29. Ross, Kenneth A.; Wright, Charles R.B. *Discrete Mathematics*. Englewood Cliffs, New Jersey: Prentice-Hall; 1985; ISBN 0-13-215286-X.

A large book certainly suitable for a two-semester course for freshmen or sophomores in computer science or mathematics. Considerable attention is paid to algorithms, but the approach is generally that of a mathematician rather than a computer scientist. Contents: Introduction to graphs and trees; sets; elementary logic and induction; functions and sequences; matrices and other semigroups; counting; more logic and induction; relations; graphs; trees; boolean algebra; algebraic systems.

30. Sahni, Sartaj. *Concepts in Discrete Mathematics.* Fridley, Minn.: Camelot Publishing Company; 1981; ISBN 0-942450-00-0.

    The author says the book is for students of computer science and engineering, with a bias towards the former, and contains needed topics not included in typical calculus and algebra courses. Algorithmic in flavor, but moderately formal. Probably a year course at the sophomore-junior level. Many interesting examples not done elsewhere. Contents: Logic; constructive proofs and mathematical induction; sets; relations; functions, recursion, and computability; analysis of algorithms; recurrence relations; combinatorics and discrete probability; graphs; modern algebra.

31. Sedlock, James T. *Mathematics for Computer Studies.* Belmont, Calif.: Wadsworth Publishing Company; 1985; ISBN 0-534-04326-7.

    Intended as first college mathematics course for computer science majors. Unsophisticated, at the level of finite mathematics, without proofs or rigor. Contents: Introduction; computer-related arithmetic; sets; combinatorics, and probability; computer-related logic; computer-related linear mathematics; selected topics (mathematics of finance, statistics, functions, induction); introduction to advanced topics (graphs and trees, semigroups, finite-state machines, languages and grammars).

32. Skvarcius, Romualdas; Robinson, William. *Discrete Mathematics with Computer Science Applications.* Menlo Park, Calif.: Benjamin Cummings Publishing Company; 1986; ISBN 0-8053-7044-7.

    "...intended audience is freshmen and sophomore students who are taking a concentration in computer science ...". Contents: Introduction to discrete mathematics; logic and sets; relations and functions; combinatorics; undirected graphs; directed graphs; boolean algebra; algebraic systems; machines and computations; probability.

33. Stanat, Donald F.; McAllister, David F. *Discrete Mathematics in Computer Science.* Englewood Cliffs, New Jersey: Prentice-Hall; 1977; ISBN 0-13-216150-8.

    A sophomore-junior level course; students will need some previous exposure to college-level mathematics. The first discrete mathematics text to consider program verification in its coverage of mathematical reasoning. The text is essentially non-algorithmic, but contains a special section on analysis of searching and sorting algorithms. Contents: Mathematical models; mathematical reasoning; sets; binary relations; functions; counting and algorithm analysis; infinite sets; algebras.

34. Tremblay, Jean-Paul; Manohar, Ram. *Discrete Mathematical Structures with Applications to Computer Science.* New York: McGraw-Hill; 1975;

ISBN 0-07-065142-6.

    An interesting feature of this text is that its first hundred pages are devoted to logic. All in all, solid coverage of the standard material. Boolean algebra is treated as a subclass of lattices. The notation for algorithms can become forbidding—see page 266 in particular. Contents: Mathematical logic; set theory; algebraic structures; lattices and boolean algebra; graph theory; introduction to computability theory.

35. Tucker, Alan C. *Applied Combinatorics, Second Edition.* New York: John Wiley and Sons; 1984; ISBN 0-471-86371-8.

    A text suitable for a wide range of audiences, from sophomores to graduate students. Contents: Elements of graph theory; covering circuits and graph coloring; trees and selections; generating functions; recurrence relations; inclusion-exclusion; Pólya's enumeration formula; combinatorial modeling in theoretical computer science; games with graphs; appendix on set theory and logic, mathematical induction, probability, the pigeonhole principle, and *Mastermind.*

References

This bibliography lists some of the materials that the planner or instructor of a lower-division discrete mathematics course might want to consult. It includes books that may be too advanced or too specialized to belong in the BIBLIOGRAPHY: TEXTBOOKS. It also lists reports and journal articles more or less pertinent to the theme.

1. ACM/IEEE Computer Society Joint Task Force. "Computer science program requirements and accreditation." *Communications of the ACM,* 1984, 27 (4) 330-335.

2. ACM Curriculum Committee on Computer Science. "Curriculum '78: Recommendations for academic programs in computer science." *Communications of the ACM,* 1979, 22 (3) 47-166.

3. Aho, A.; Hopcroft, J.; Ullman, J. *Data Structures and Algorithms.* Reading, Mass.: Addison-Wesley, 1983.

    Primarily a book on data structures. Algorithms are presented in Pascal. An accessible reference for analysis of algorithms.

4. Bavel, Zamir. *Math Companion for Computer Science.* Reston, Virginia: Reston Publishing Company, 1982, ISBN 0-8359-4300-3 or 0-8359-4299-6 (pbk).

    A manual, not a text.

5. Beidler, John; Austing, Richard H.; Cassel, Lillian N. "Computing programs in small colleges." *Communications of the ACM,* 1985, 28 (6) 605-611.

    Summary report of The ACM Small College Task Force; outlines resources, courses, and problems for small colleges developing degree programs in computing. See especially "The Mathematics Component," p. 610.

6. Bellman, Richard; Cooke, Kenneth; Lockett, Jo Ann. *Algorithms, Graphs, and Computers.* New York: Academic Press, 1970, ISBN 0-12-084840-6.

7. Berztiss, A.T. *Data Structures: Theory and Practice, Second Edition.* New York: Academic Press, 1975, ISBN 0-12-093552-X.
   Although dated, primarily by its dependence on FORTRAN, can be used as a reference on representation of digraphs by trees, and on critical path analysis.

8. Berztiss, Alfs T. *Towards a Rigorous Curriculum for Computer Science.* Technical Report 83-5, University of Pittsburgh Department of Computer Science, 1983.

9. Bogart, K.P.; Cordiero, K.; Walsh, M.L. "What is a discrete mathematics course?" *SIAM News,* 1985, 18(1).

10. Deo, Narsingh. *Graph Theory with Applications to Engineering and Computer Science.* Englewood Cliffs, New Jersey: Prentice-Hall, 1974, ISBN 0-13-363473-6.
    An excellent source for applications of the theory of graphs, both undirected and directed, but somewhat dated. Contains extensive bibliographies (to 1972).

11. Dornhoff, Larry L.; Hohn, Franz E. *Applied Modern Algebra.* New York: Macmillan Publishing Company, 1978, ISBN 0-02-329980-0.
    Deals in great detail with applications of algebra in computer engineering, but the complicated notation makes access to the application studies rather hard.

12. Gibbs, Norman E.; Tucker, Allen B. "Model curriculum for a liberal arts degree in computer science." *Communications of the ACM,* 1986, 29 (3).
    A curriculum developed by computer scientists supported by a grant from the Sloan Foundation. The purpose was to define a rigorous undergraduate major in computer science for liberal arts colleges.

13. Gordon, Sheldon P. "A discrete approach to the calculus." *Int. J. Math. Educ. Sci. Technol.,* 1979, 10 (1) 21-31.
    A report on an experiment in using discrete calculus (finite differences and sums in place of derivatives and integrals) as an introduction to computer-based continuous calculus.

14. Gries, David. *The Science of Programming.* New York: Springer-Verlag, 1981, ISBN 0-387-90641-X.
    This book deals with the development of correct programs. The introductory sections on logic achieve thoroughness without becoming intimidating. Apart from these sections there is little relevance to the proposed discrete mathematics course.

15. Harary, Frank. *Graph Theory.* Reading, Mass.: Addison-Wesley, 1969, ISBN 0-201-02787-9.
    Deals almost exclusively with undirected graphs. Extensive bibliography (to 1968). Heavy emphasis on enumeration. Most exercises require considerable mathematical maturity.

16. Hart, Eric W. "Is discrete mathematics the new math of the eighties?" *Mathematics Teacher,* 1985, 75 (5) 334-338.

17. Heid, M. Kathleen. "Calculus with muMath: Implications for curriculum reform." *The Computing Teacher,* 1983, 11 (4) 46-49.

18. Hodgson, Bernard R.; Poland, John. "Revamping the mathematics curriculum: The influence of computers." *Notes of the Canadian Mathematical Society,* 1983, 15 (8) n.p.
    A product of a meeting of a working group of the Canadian Mathematics Education Study Group.

19. IEEE Educational Activities Board, Model Program Committee. *The 1983 IEEE Computer Society Model Program in Computer Science and Engineering.* New York: The Institute of Electrical and Electronics Engineers, 1983.
    A section "Discrete Mathematics" appears on pp. 8-12. This contains a rather demanding modular outline of topics to be covered, and urges integration of the mathematical theory with computer science and engineering applications.

20. Knuth, Donald E. *The Art of Computer Programming, Volume 1: Fundamental Algorithms.* Reading, Mass.: Addison-Wesley, 1973, ISBN 0-201-03809-9.
    Still the standard reference on the tools for analysis of algorithms.

21. Koffman, E.; Miller, P.; Wardle, C. "Recommended curriculum for CS1, 1984: A report of the ACM curriculum committee task force for CS1." *Communications of the ACM,* 1985, 27 (10) 998-1001.

22. Koffman, E.; Stemple, D.; Wardle, C. "Recommended curriculum for CS2, 1984—A report of the ACM curriculum committee task force for CS2." *Communications of the ACM,* 1985, 28 (8) 815-818.

23. Laufer, Henry. *Applied Modern Algebra.* Boston: Prindle, Weber, and Schmidt, 1984, ISBN 0-87150-702-1.
    Sophomore-junior level; intended to follow the description for an applied algebra course in the 1981 MAA-CUPM recommendations (see item 25 below).

24. Lidl, Rudolf; Pilz, Gunter. *Applied Abstract Algebra.* New York: Springer-Verlag, 1985, ISBN 0-387-96035-X.
    Describes aspects of abstract algebra applicable to discrete mathematics.

25. MAA Committee on the Undergraduate Program

in Mathematics. *Recommendations for a General Mathematical Sciences Program.* Washington, DC: The Mathematical Association of America, 1981.

26. Manna, Zohar; Waldinger, Richard. *The Logical Basis for Computer Programming, Volume 1: Deductive Reasoning.* Reading, Mass.: Addison-Wesley, 1985, ISBN 0-201-18260.

    This book is in two parts. The first is a thorough exposition of logic. In the second, data types are treated as theories. Although the authors see the material as ultimately replacing calculus, it probably could not do so as presented—the departure from established practice is too radical, and the style is too austere for an undergraduate text.

27. Maurer, Stephen B. "The algorithmic way of life is best." *The College Mathematics Journal,* 1985, 16 (1) 2-5.

    This is the lead article for a "forum" in which others contribute their own very diverse points of view on the general theme. The whole exchange occupies pp. 2-21.

28. Mulder, Michael C.; Dalphin, John. "Computer science program requirements and accreditation." *Computer,* 1984, 17 (4) 30-35.

29. Nijenhuis, Albert; Wilf, Herbert S. *Combinatorial Algorithms.* New York: Academic Press, 1975, ISBN 0-12-519250-9.

    A study of aspects of computers, algorithms, and mathematics, and of the relations between them, based on an examination of programs in FORTRAN.

30. Preparata, Franco P.; Yeh, Raymond T. *Introduction to Discrete Structures for Computer Science and Engineering.* Reading, Mass.: Addison-Wesley, 1973, ISBN 0-201-05968-1.

    Junior level. Boolean algebras are arrived at via lattices. Solid coverage, but the material on graphs needs to be supplemented.

31. Ralston, A. "Computer science, mathematics, and the undergraduate curricula in both." *American Mathematical Monthly,* 1981, 88 (7) 472-485.

32. Ralston, A. "The really new college mathematics and its impact on the high school curriculum." In Hirsch, Christian R.; Zweng, Marilyn J. (Eds.), *The Secondary School Mathematics Curriculum: 1985 Yearbook.* Reston, Virginia: National Council of Teachers of Mathematics, 1985, 29-42, ISBN 0-87353-217-1.

33. Ralston, A. "The first course in computer science needs a mathematical corequisite." *Communications of the ACM,* 1984, 27 (10) 1002-1005.

34. Ralston, A.; Shaw, M. "Curriculum '78—Is computer science really that unmathematical?" *Com-munications of the ACM,* 1980, 23 (2) 67-70.

35. Ralston, A.; Young, G.S. (Eds.). *The Future of College Mathematics.* New York: Springer-Verlag, 1983, ISBN 0-387-90813-7.

    Proceedings of the 1982 "Williams Conference" that added impetus to the movement to add more discrete mathematics to the lower-division curriculum. The ideas presented in the book are far from exhausted.

36. Reingold, Edward M.; Nievergelt, Jurg; Deo, Narsingh. *Combinatorial Algorithms: Theory and Practice.* Englewood Cliffs, New Jersey: Prentice-Hall, 1977, ISBN 0-13-15244-7.

37. Roberts, Fred S. *Applied Combinatorics.* Englewood Cliffs, New Jersey: Prentice-Hall, 1984, ISBN 0-13-039313-4.

    Junior-senior level; wide range of applications; algorithms.

38. Roberts, Fred S. *Discrete Mathematical Models with Applications to Social, Biological, and Environmental Models.* Englewood Cliffs, New Jersey: Prentice-Hall, 1976, ISBN 0-13-214171-X.

39. Shaw, M. (Ed.). *The Carnegie-Mellon Curriculum for Undergraduate Computer Science.* New York: Springer-Verlag, 1984.

40. Smith, Douglas; Eggen, Maurice; St. Andre, Richard. *A Transition to Advanced Mathematics.* Monterey, Calif.: Brooks-Cole, 1983, ISBN 0-534-01249-3.

    This book is intended to pave the way from calculus to more advanced mathematics. Its core chapters deal with logic, proofs, sets, relations, functions, and cardinality.

41. Solow, Daniel. *How To Read and Do Proofs.* New York: John Wiley and Sons, 1982, ISBN 0-471-86645-8.

    For use as a supplement if the discrete mathematics course emphasizes proof.

42. Wand, Mitchell. *Induction, Recursion, and Programming.* New York: North Holland, 1980, ISBN 0-444-00322-3.

    This text is addressed specifically to computer science students. It is based on the thesis that a program is a mathematical object. Correctness of programs is then established by reasoning about these mathematical objects in an appropriate language. Prior exposure to sets and functions is assumed.

43. Whitehead, Earl Glen, Jr. *Combinatorial Algorithms.* New York: Courant Institute of Mathematical Sciences, New York University, 1973.

44. Wirth, Niklaus. *Algorithms and Data Structures.* Englewood Cliffs, New Jersey: Prentice-Hall, 1986, ISBN 0-13-022005-1.

Calculus Transition: From High School to College

From 1983 to 1986, the CUPM Panel on Calculus Articulation studied problems associated with college transition for students who had studied calculus in high school. The report of the CUPM Panel originally appeared under the title "Transition from High School to College Calculus" in the AMERICAN MATHEMATICAL MONTHLY, *October, 1987. It is reprinted here with minor editorial changes.*

1989 Preface

The importance of the problems identified in the panel's report has been underscored by several recent international and national assessments of mathematics education (e.g., The National Research Council's report *Everybody Counts: A Report to the Nation on the Future of Mathematics Education*).

Since the panel's report was written in 1986, the severity of the transition problems from high school to college calculus has increased both qualitatively and quantitatively. In 1987 there were 59,123 students who took an Advanced Placement Calculus Examination, an increase of 85% from 1982. Although this increase does not necessarily indicate a similar growth rate in the number of students studying calculus in high school, it does document a large increase in the number of students entering college calculus having earned an Advanced Placement Calculus score of three or less. This increase serves to intensify the Report's recommendation that colleges and universities develop special calculus courses for these students.

The development of Computer Algebra Systems designed for classroom teaching introduces a new component into the transition problems students may encounter in going from high school to college calculus. It is vitally important to expand communication between college and high school teachers in regards to the development of this technology, particularly with respect to pedagogical issues.

Donald B. Small
Colby College
March, 1989

Introduction

There is a widespread and growing dissatisfaction with the performance in college calculus courses of many students who had studied calculus in high school. In response to this concern, in the fall of 1983, the Committee on the Undergraduate Program in Mathematics (CUPM) formed a Panel on Calculus Articulation to undertake a three-year study of questions concerning the transition of students from high school calculus to college calculus and submit a report to CUPM detailing the problems encountered and proposals for their solution.

The seriousness of the issues involved in the Panel's study is underscored by the number of students involved and their academic ability. During the ten-year period 1973 to 1982, the number of students in high school calculus courses grew at a rate exceeding 10% annually. Of the 234,000 students who passed a high school calculus course in 1982, 148,600 received a grade of B- or higher [2]. Assuming a continuation of the 10% growth rate and a similar grade distribution there were approximately 200,000 high school students in the spring of 1985 who received a grade of B- or higher in a calculus course. Thus possibly a third or more of the 500,000 college students who began their college calculus program (in Calculus I, Calculus II, or Calculus III) in the fall of 1985 had already received a grade of B- or higher in a high school calculus course.

The students studying calculus in high school constitute a large majority of the more mathematically capable high school students. (In 1982, 55% of high school students attended schools where calculus was taught [2].) Students who score a 4 or 5 on an Advanced Placement (AP) Calculus examination normally do well in maintaining their accelerated mathematics program during the transition from high school to college. However, this is a very small percentage of the students who take calculus in high school. For example, in 1982, of the 32,000 students who took an Advanced Placement calculus examination, just over 12,000 received scores of 4 or 5, which represents only 6% of all high school students who took calculus that year. The primary concern of the Panel was with the transition difficulties associated with the remaining almost 94% of the high school calculus students.

Problem Areas

Past studies and the Panel's surveys of high school teachers, college teachers, and state supervisors suggest that the major problems associated with the transition from high school calculus to college calculus are:

1. High school teacher qualifications and expectations.
2. Student qualifications and expectations.
3. The effect of repeating a course in college after having experienced success in a similar high school course.
4. College placement.
5. Lack of communication between high schools and colleges.

These problems were addressed by first considering accelerated programs in general, high school calculus (successful, unsuccessful), and the responsibilities of the colleges.

Accelerated Programs

Accelerated mathematics programs, usually beginning with algebra in eighth grade, are now well established and accepted in most school systems. The success of these programs in attracting the mathematically capable students was documented in the 1981-82 testing that was done for the "Second International Mathematics Study." The Summary Report [9] states with reference to a comparison between twelfth grade precalculus students and twelfth grade calculus students in the United States:

> We note furthermore that in every content area (sets and relations, number systems, algebra, geometry, elementary functions/calculus, probability and statistics, finite mathematics), the end-of-the-year average achievement of the precalculus classes was less (and in many cases considerably less) than the beginning-of-the-year achievement of the calculus students.

The report continues:

> It is important to observe that the great majority of U.S. senior high school students in fourth and fifth year mathematics classes (that is, those in precalculus classes) had an average performance level that was at or below that of the lower 25% of the countries. The end-of-year performance of the students in the calculus classes was at or near the international means for the various content areas, with the exception of geometry. Here U.S. performance was below the international average.

Thus those students in accelerated programs culminating in a calculus course perform near the international mean level while their classmates in (non-accelerated) programs culminating in a precalculus

course perform in the bottom 25% in this international survey. The poor performance in geometry by both the precalculus and calculus students correlates well with the statistic that 38% of the students were never taught the material contained in the geometry section of the test [9, p. 59]. The test data underscores the concern expressed by many college teachers that more emphasis needs to be placed on geometry throughout the high school curriculum. This data does not, however, indicate that accelerated programs emphasize geometry less than non-accelerated programs.

The success of the accelerated programs in completing the "normal" four year high school mathematics program by the end of the eleventh grade presents schools with both an opportunity and a challenge for a "fifth" year program. There are two acceptable options:

1. Offer college-level mathematics courses that would continue the students' accelerated program and thus provide exemption from one or two semesters of college mathematics.
2. Offer high school mathematics courses that would broaden and strengthen a student's background and understanding of precollege mathematics.

Not offering a fifth year course or offering a watered-down college level course with no expectation of students earning advanced placement are not considered to be acceptable options.

A great deal of prestige is associated with offering calculus as a fifth year course. Communities often view the offering of calculus in their high school as an indication of a quality educational program. Parents, school board officials, counselors, and school administrators often demonstrate a competitive pride in their school's offering of calculus. This prestige factor can easily manifest itself in strong political pressure for a school to offer calculus without sufficient regard to the qualifications of teachers or students.

It is important that this political pressure be resisted and that the choice of a fifth year program be made by the mathematics faculty of the local school and be made on the basis of the interest and qualifications of the mathematics faculty and the quality and number of accelerated students. School officials should be encouraged to develop public awareness programs to extend the prestige and support that exists for the calculus to acceleration programs in general. This would help diffuse the political pressure as well as broaden school support within the community.

Schools that elect the first option of offering a college level course should follow a standard college course syllabus (e.g., the Advanced Placement syllabus for calculus). They should use placement test scores along with the college records of their graduates as primary measures of the validity of their course.

For schools that elect the second option, a variety of courses is possible. The following course descriptions represent four possibilities.

ANALYTICAL GEOMETRY. This course could go well beyond the material normally included in second year algebra and precalculus. It could include Cartesian and vector geometry in two- and three-dimensions with topics such as translation and rotation of axes, characteristics of general quadratic relations, curve sketching, polar coordinates, and lines, planes, and surfaces in three-dimensional space. Such a course would provide specific preparation for calculus and linear algebra, as well as give considerable additional practice in trigonometry and algebraic manipulations.

PROBABILITY AND STATISTICS. This course could be taught at a variety of levels, to be accessible to most students, or to challenge the strongest ones. It could cover counting methods and some topics in discrete probability such as expected values, conditional probability, and binomial distributions. The statistics portion of the course could emphasize exploratory data analysis including random sampling and sampling distributions, experimental design, measurement theory, measures of central tendency and spread, measures of association, confidence intervals, and significance testing. Such an introduction to probability and statistics would be valuable to all students, and for those who do not plan to study mathematics, engineering, or the physical sciences, probably more valuable than a calculus course.

DISCRETE MATHEMATICS. This type of course could include introductions to a number of topics that are either ignored or treated lightly within a standard high school curriculum, but which would be stimulating and widely useful for the college-bound high school student. Suggested topics include permutations, combinations, and other counting techniques: mathematical induction; difference equations; some discrete probability; elementary number theory and modular arithmetic; vector and matrix algebra, perhaps with an introduction to linear or dynamic programming; and graph theory.

MATRIX ALGEBRA. This course could include basic arithmetic operations on matrices, techniques for finding matrix inverses, and solving systems of linear equations and their equivalent matrix equations using Gaussian elimination. In addition, some introduction to linear programming and dynamic programming could be included. This course could also emphasize three-dimensional geometry.

High School Calculus

There are many valid reasons why a fifth year program should include a calculus course. Four major reasons: (1) calculus is generally recognized as the starting point of a college mathematics program, (2) there exists a (nationally accepted) syllabus, (3) the Advanced Placement program offers a nation-wide mechanism for obtaining advanced placement, and (4) there is a large prestige factor associated with offering calculus in high school. Calculus, however, should not be offered unless there is a strong indication that the course will be successful.

Successful Calculus Courses

The primary characteristics of a successful high school calculus course are:

1. A qualified and motivated instructor with a mathematics degree that included at least one semester of a junior-senior real analysis course involving a rigorous treatment of limits, continuity, etc.
2. Administrative support, including provision of additional preparation time for the instructor (e.g., as recommended by the North Central Accreditation Association).
3. A full year program based on the Advanced Placement syllabus.
4. A college text should be used (not a watered-down high school version).
5. Advanced placement for students (rather than mere preparation for repeating calculus in college) is a major goal.
6. Course evaluation based primarily on college placement and the performance of its graduates in the next higher level calculus course.
7. Restriction of course enrollment to only qualified and interested students.
8. The existence of an alternative fifth year course that students may select who are not qualified for or interested in continuing in an accelerated program.

The bottom line of what makes a high school calculus course successful is no surprise to anyone. A qualified teacher with high but realistic expectations, using somewhat standard course objectives, and students who are willing and able to learn result in a successful transition at any level of our educational process. Problems appear when any of the above ingredients are missing.

Unsuccessful Calculus Courses

Two types of high school calculus courses have an

undesirable impact on students who later take calculus in college.

One type is a one semester or partial year course that presents the highlights of calculus, including an intuitive look at the main concepts and a few applications, and makes no pretense about being a complete course in the subject. The motivation for offering a course of this kind is the misguided idea that it prepares students for a *real* course in college.

However, such a preview covers only the glory and thus takes the excitement of calculus away from the college course without adequately preparing students for the hard work and occasional drudgery needed to understand concepts and master technical skills. Professor Sherbert has commented: "It is like showing a ten minute highlights film of a baseball game, including the final score, and then forcing the viewer to watch the entire game from the beginning—with a quiz after each inning."

The second type of course is a year-long, semiserious, but watered-down treatment of calculus that does not deal in depth with the concepts, covers no proofs or rigorous derivations, and mostly stresses mechanics. The lack of both high standards and emphasis on understanding dangerously misleads students into thinking they know more than they really do.

In this case, not only is the excitement taken away, but an unfounded feeling of subject mastery is fostered that can lead to serious problems in college calculus courses. Students can receive respectable grades in a course of this type, yet have only a slight chance of passing an examination. Those who place into second term calculus in college will find themselves in heavy competition with better prepared classmates. Those who elect (or are selected) to repeat first term calculus believe they know more than they do, and the motivation and willingness to learn the subject are lacking.

College Programs

Several studies ([1], [3], [5], [6], [7]) have been conducted on the performance in later courses by students who have received advanced placement (and possibly college credit) by virtue of their scores on Advanced Placement Calculus examinations. The studies show that, overall, students earning a score of 4 or 5 on either the AB or BC Advanced Placement Calculus examination do as well or better in subsequent calculus courses than the students who have taken all their calculus in college. It is therefore strongly recommended that colleges recognize the validity of the Advanced Placement

Calculus program by the granting of one semester advanced placement with credit in calculus for students with a 4 or 5 score on the AB examination, and two semesters of advanced placement with credit in calculus for students with a 4 or 5 score on the BC examination.

The studies reviewed by the Panel do not indicate any clear conclusions concerning performance in subsequent calculus courses by students who have scored a 3 on an Advanced Placement Calculus examination. The treatment of these students is a very important transition problem since approximately one-third of all students who take an Advanced Placement Calculus examination are in this group and many of them are quite mathematically capable.

It is therefore recommended that these students be treated on a special basis in a manner that is appropriate for the institution involved. For example, several colleges offer a student who has earned a 3 on an Advanced Placement Calculus examination the opportunity to upgrade this score to an "equivalent 4" by doing sufficiently well on a Department of Mathematics placement examination. Another option is to give such students one semester of advanced placement with credit for Calculus I upon successful completion of Calculus II. A third option is to give one semester of advanced placement with credit for Calculus I and provide a special section of Calculus II for such students.

Other important transition problems are associated with students who have studied calculus in high school, but have not attained advanced placement either through the Advanced Placement Calculus program or effective college procedures. These students pose an important and difficult challenge to college mathematics departments, namely: How should these students be dealt with so that they can benefit from their accelerated high school program and not succumb to the negative and (academically) destructive attitude problems that often result when a student repeats a course in which success has already been experienced? There are three major factors to consider with respect to these students.

1. The lack of uniformity of high school calculus courses. The wide diversity in the backgrounds of the students necessitates that a large review component be included in their first college calculus course to guarantee the necessary foundation for future courses.

2. The mistaken belief of most of these students that they really know the calculus when, in fact, they do not. Thus they fail to study enough at the beginning of the course. When they realize their mistake (if they do), it is often too late. These students often

become discouraged and resentful as a result of their poor performance in college calculus, and believe that it is the college course that must be at fault.

3. The "Pecking Order" syndrome. The better the student, the more upsetting are the understandable feelings of uncertainty about his or her position relative to the others in the class. Although this is a common problem for all college freshmen, it is compounded when the student appears to be repeating a course in which success had been achieved the preceding year. This promotes feelings of anxiety and produces an accompanying set of excuses if the student does not do at least as well as in the previous year.

The uncertainty of one's position relative to the rest of the class often manifests itself in the student not asking questions or discussing in (or out of) class for fear of appearing *dumb*. This is in marked contrast to the highly confident high school senior whose questions and discussions were major components in his or her learning process.

The unpleasant fact is that the majority of students who have taken calculus in high school and have not clearly earned advanced placement do not *fit* in either the standard Calculus I or Calculus II course. The students do not have the level of mastery of Calculus I topics to be successful if placed in Calculus II and are often doomed by attitude problems if placed in Calculus I. In modern parlance, this is the *rock and hard place*.

An additional factor to consider is the negative effect that a group of students who are repeating most of the content of Calculus I has on the rest of the class as well as on the level of the instructor's presentations.

What is needed are courses designed especially for students who have taken calculus in high school and have not clearly earned advanced placement. These courses need to be designed so that they:

1. Acknowledge and build on the high school experiences of the students;
2. Provide necessary review opportunities to ensure an acceptable level of understanding of Calculus I topics;
3. Are *clearly different* from high school calculus courses (in order that students do not feel that they are essentially just repeating their high school course);
4. Result in an equivalent of one semester advanced placement.

Altering the traditional lecture format or rearranging and supplementing content seem to be two promising approaches to developing courses that will satisfy the above criteria. For example, Colby College has successfully developed a two semester calculus course that fulfills the four conditions. The course integrates multivariable with single variable calculus, and thereby covers the traditional three semester program in two semesters [10].

Of course, the introduction of a new course entails an accompanying modification of college placement programs. However, providing new or alternative courses should have the effect of simplifying placement issues and easing transition difficulties that now exist.

Recommendations

1. School administrators should develop public awareness programs with the objective of extending the support that exists for fifth year calculus courses to accelerated programs including all of the fifth year options.
2. A fifth year program should offer a student a choice of courses (not just calculus).
3. The choice of fifth year options should be made by the high school mathematics faculty on the basis of their interest and qualifications and the quality and number of the accelerated students.
4. If a fifth year course is intended as a college level course, then it should be treated as a college level course (text, syllabus, rigor).
5. A fifth year college level course should be taught with the expectation that successful graduates (B- or better) would not repeat the course in college.
6. A fifth year program should provide an alternative option for the student who is not qualified to continue in an accelerated program.
7. A mathematics degree that includes at least one semester of a junior-senior real analysis course involving a rigorous treatment of limit, continuity, etc., is strongly recommended for anyone teaching calculus.
8. A high school calculus course should be a full year course based on the Advanced Placement syllabus.
9. The instructor of a high school calculus course should be provided with additional preparation time for this course.
10. High school calculus students should take either the AB or BC Advanced Placement calculus examination.
11. The evaluation of a high school calculus course should be based primarily on college placement and the performance of its graduates in the next level calculus course.

12. Only interested students who have successfully completed the standard four year college preparatory program in mathematics should be permitted to take a high school calculus course.

13. Colleges should grant credit and advanced placement out of Calculus I for students with a 4 or 5 score on the AB Advanced Placement calculus examination, and credit and advanced placement out of Calculus II for students with a 4 or 5 score on the BC Advanced Placement calculus examination. Colleges should develop procedures for providing special treatment for students who have earned a score of 3 on an Advanced Placement calculus examination.

14. Colleges should individualize as much as possible the advising and placement of students who have taken calculus in high school. Placement test scores and personal interviews should be used in determining the placement of these students.

15. Colleges should develop special courses in calculus for students who have been successful in accelerated programs, but have clearly not earned advanced placement.

Colleges have an opportunity and responsibility to develop and foster communication with high schools. In particular:

16. Colleges should establish periodic meetings where high school and college teachers can discuss expectations, requirements, and student performance.

17. Colleges should coordinate the development of enrichment programs (courses, workshops, institutes) for high school teachers in conjunction with school districts and state mathematics coordinators.

References

[1] C. Cahow, N. Christensen, J. Gregg, E. Nathans, H. Strobel, G. Williams. Undergraduate Faculty Council of Arts and Sciences Committee on Curriculum; Subcommittee on Advanced Placement Report, Trinity College, Duke University, 1979.

[2] C. Dennis Carroll. "High school and beyond tabulation: Mathematics courses taken by 1980 high school sophomores who graduated in 1982." National Council of Education Statistics, April 1984 (LSB 84-4-3).

[3] P.C. Chamberlain, R.C. Pugh, J. Schellhammer. "Does advanced placement continue throughout the undergraduate years?" College and University, Winter 1968.

[4] "Advanced Placement Course Description, Mathematics." The College Board, 1984.

[5] E. Dickey. "A study comparing advanced placement and first-year college calculus students on a calculus achievement test." Ed.D. dissertation, University of South Carolina, 1982.

[6] D.A. Frisbie. "Comparison of course performance of AP and non-AP calculus students." Research Memorandum No. 207, University of Illinois, September 1980.

[7] D. Fry. "A comparison of the college performance in calculus-level mathematics courses between regular-progress students and advanced placement students." Ed.D. dissertation, Temple University, 1973.

[8] C. Jones, J. Kenelly, D. Kreider. "The advanced placement program in mathematics—Update 1975." Mathematics Teacher, 1975.

[9] Second International Mathematics Study Summary Report for the United States. Champaign, IL: Stipes Publishing, 1985.

[10] D. Small, J. Hosack. Calculus: An Integrated Approach. McGraw-Hill, 1990.

[11] D.H. Sorge, G.H. Wheatley. "Calculus in high school—At what cost?" American Mathematical Monthly 84 (1977) 644-647.

[12] D.M. Spresser. "Placement of the first college course." International Journal Mathematics Education, Science, and Technology 10 (1979) 593-600.

Panel Members

DONALD B. SMALL, CHAIR, Colby College.

GORDON BUSHAW, Central Kitsap High School, Silverdale, Washington.

JOHN H. HODGES, University of Colorado.

DONALD J. NUTTER, Firestone High School, Akron, Ohio.

RONALD SCHNACKENBURG, Steamboat Springs High School, Colorado.

DONALD R. SHERBERT, University of Illinois.

BARBARA STOTT, Riverdale High School, Jefferson, Louisiana.

Curriculum for Grades 11-13

The following report was approved in 1987 by the Board of Directors of the National Council of Teachers of Mathematics and by the Board of Governors of the Mathematical Association of America. The joint NCTM-MAA committee that prepared the report was chaired by Joan Leitzel, Associate Provost of The Ohio State University. It is printed here in its entirety for the first time.

The Joint Task Force on Curriculum for Grades 11-13 was formed in Spring 1986 by John Dossey, President of the National Council of Teachers of Mathematics, and Lynn Steen, President of the Mathematical Association of America. The Task Force was charged to focus on curriculum for the mainstream of students who take a college preparatory program in high school and who go on in college to a standard freshman mathematics course such as calculus, finite mathematics, statistics, or discrete mathematics. The Task Force was asked to advise on the need for new recommendations to high schools and colleges concerning curriculum for students who intend to pursue careers that depend on mathematics.

The Task Force has studied and synthesized recent national reports on the state of mathematics education, as well as the recommendations of many national and state boards, and the reports of several recent curriculum projects. It is clear that the present situation in mathematics education is dynamic and that significant changes in curriculum in grades 11-13 may occur in the next few years. The Task Force has been impressed and excited by the quality and scope of curricular projects currently underway. We note particularly the K-12 Curriculum Project undertaken by the Mathematical Sciences Education Board (MSEB), the frameworks of which will be released during 1989, and the collegiate mathematics project undertaken jointly by MSEB and the Board of Mathematical Sciences of the National Research Council (The Mathematical Sciences in the Year 2000: Assessment for Renewal in U.S. Colleges and Universities).

It is not the purpose of this report to resolve present conflicts or to determine new directions for mathematics education. Rather, it is the purpose of this report to summarize those areas related to grades 11-13 curriculum in which it appears the profession now has a level of consensus, and also to summarize those issues where there is lack of consensus and apparent need of additional study.

Although it is tempting to make no statements during this time of rapid evolution, we are persuaded that school and university students, parents, and teachers will welcome some clarification from the professional associations. The joint statement of the NCTM and MAA on college preparation has not been updated since it was released in the mid-70s. Thus, although we do not in any way want to suggest that the status quo is acceptable for education in mathematics, we do wish to summarize prevailing opinions in the hope that some guidance will be helpful to students and teachers at this time.

This Task Force was formed to consider curricular matters. Even though it is beyond the charge of the Task Force, we are compelled to mention that attention to curriculum will be meaningless without concurrent attention to other matters. One such concern is the increasing shortage of qualified teachers at all levels, especially in the middle grades and the secondary schools. In addition, colleges and universities frequently depend heavily on teaching assistants and part-time teachers.

Secondly, it is difficult to discuss curriculum in grades 11-13 without reference to the mathematics of the earlier grades. The Task Force has been reminded frequently of the critical role that middle school mathematics and the first year of algebra play in a student's success in college preparatory mathematics. Continued attention needs to be given to the curriculum of these grades. Indeed, some of the recommendations we make in this report for grades 11-13 apply as well to earlier grades.

Issues of Apparent Consensus

After reviewing numerous reports and talking with teachers, mathematicians, and mathematics educators, the Task Force believes that there is agreement within the profession on many issues related to curriculum in grades 11-13 for those students who are directed toward mathematics-dependent collegiate programs. From this consensus, it appears that the following recommendations can be made:

1. College-bound students should take mathematics in all years of secondary school; this mathematics should include the content of intermediate algebra and geometry.

2. Geometry in both two and three dimensions, coordinate geometry and the development of geometric perception are essential parts of college preparatory mathematics.

3. Although computer programming is an important tool for all college-bound students, courses in computer programming should not be regarded as substitutes for college preparatory mathematics.

4. Students should learn the content of a full four years of college preparatory mathematics before taking a calculus course. Students directed toward twelfth grade calculus need an enriched program as early as grades 7 and 8.

5. A full program of college preparatory mathematics should be provided in every secondary school.

6. Calculus, if taught in the schools, should be equivalent to college-level calculus. NCTM and MAA have made a joint recommendation to the schools that students should expect to write proficiency examinations (either Advanced Placement or university credit exams) to establish that they have learned beginning calculus, and to obtain college credit for high school calculus courses. Watered-down calculus courses in the secondary schools stressing manipulations but slighting subtle processes do not help students. Introducing polynomial calculus in pre-calculus courses uses time better spent on other topics.

7. If mathematics is required in a student's college program, the student should enroll in mathematics courses as a college freshman. There should not be a gap of a year or more during which a student takes no mathematics.

8. Students should expect to make use of calculators and computers in their mathematics courses in grades 11-13. Calculators and computers should be used as tools to enhance and expand the learning of mathematics in these grades.

9. The availability of technology should permit computational approaches to college preparatory mathematics that result in this mathematics being accessible to more students. Further, the availability of technology should permit students in these grades to investigate problem situations that are not approachable without computational tools, and to be introduced to different mathematics than was possible without calculators and computers.

10. Mathematics in grades 11-13 needs to have a clear connection to real world problems, and students should be expected to acquire a growing ability to use mathematics to model real situations. Students should become aware of new applications of mathematics as these applications develop. In addition, students should understand that mathematics itself is a developing discipline, and, where possible, students should come to appreciate new developments in mathematics.

11. The curriculum of grades 11-13 should contain units in statistics, probability, and fundamental topics in discrete mathematics for all college-bound students.

12. Mathematics in grades 11-13 should have goals beyond the acquisition of computational techniques. Mathematical understanding and analytical reasoning are basic goals for mathematics at this level. Problem solving strategies should be stressed, and manipulative and computational techniques, although important, should not predominate.

13. Teachers of mathematics in grades 11-13 should employ strategies that encourage student reading, writing, and reflection. Assignments and examinations should be designed to help students become more independent learners of mathematics and to increase their abilities to discuss both orally and in writing the mathematical ideas they are learning. Courses should not cover an excessive number of topics at the expense of reflection and independent learning on the part of students. Teachers should take advantage of mathematics competitions and science fairs to encourage independent learning in students.

14. While there is need for meaningful review within new mathematics, the amount of time spent on review at the beginning of a course in grades 11-13 should not be excessive. Typically, review should be integrated into the learning of new mathematics.

15. Expectations of students with regard to homework, examinations, and knowledge of previous courses need to be raised in many grade 11-13 programs. These expectations should include daily homework, cumulative examinations, and examination questions that require problem-solving skills.

16. To overcome the effects of socialization that discourages girls and American minorities from studying mathematics, these groups of students should be especially encouraged in the study of mathematics in grades 11-13, and efforts should be made to identify applications of mathematics that hold particular interest for under-represented groups. Furthermore, the perceived preponderance of negative attitudes toward mathematics in this country should be studied to determine what aspects of curriculum and what features of culture contribute most heavily to these attitudes.

Recommendation

The Task Force is persuaded that improvement in the curriculum of grades 11-13 and in student performance in this curriculum requires strong collaborative effort among mathematicians, university faculty in mathematics education, teachers, school leaders, counselors, students, and parents. The Task Force recommends that funding be sought so that the curricular issues cited above and other essential related information can be communicated to these groups. Particularly targeted brochures and flyers need to be developed and circulated to teachers, school leaders, counselors, students, and parents.

We have included at the end of the report a fuller description of the rationale and content of these proposed communiques and recommendations for the desired collaboration.

Issues Requiring Further Study

In addition to the issues listed above on which the profession appears to have a consensus, we have identified many issues where there is lack of consensus. In many cases, studies, reviews, and experimentation are underway in a variety of projects. Where we know of such projects, we have cited them here. In some cases it appears that more study and discussion will be needed to clarify these issues.

Gifted Students: Acceleration or Enrichment?

The types of courses that students are capable of taking in grades 11-13 will depend upon the preparation of these students in earlier years. Should the very best students be accelerated or should their programs instead be enriched with topics they may otherwise miss? This question is still a matter of debate among mathematics educators.

Advocates of early introduction of algebra usually argue that the current curriculum for grades 7 and 8 is mainly just a review of topics taught in previous grades with little new material introduced. The Report of the MSEB Task Force on Curriculum Frameworks for K-12 Mathematics (Draft, October 1986) states that:

> We applaud the current attempts to make algebra an eighth grade subject. There is ample evidence from other countries that eighth graders can handle algebra. More generally, we think that grades 7 and 8 should look forward to high school mathematics as much or more than they look backward to elementary school mathematics.

The University of Chicago School Mathematics Program is being designed for the general school population. Its eighth grade course is mainly algebra, but

heavily manipulative techniques are postponed to later courses. In his paper, "Why Elementary Algebra Can, Should, and Must be an 8th Grade Course for Average Students," Zalman Usiskin argues that in other countries algebra is usually done with all students at grades 7 and 8 and that with proper curriculum in grades 1-6, algebra could be mastered in grade 8 by U.S. students as well.

In *The Underachieving Curriculum: Assessing U.S. School Mathematics from an International Perspective*, a report on the Second International Mathematics Study, the authors recommend:

> The content of the mathematics curriculum needs to be re-examined and revitalized. The domination of the lower secondary school curriculum by the arithmetic of the elementary school has resulted in a program that, from an international point of view, is very lean. The curriculum should be broadened and enriched by including a substantial treatment of topics such as geometry, probability, statistics and algebra, as well as promoting higher-level process goals such as estimation and problem-solving.

In "Let's Not Teach Algebra to Eighth Graders!" (*Mathematics Teacher*, November 1985), Fernand Prevost provides evidence that offering algebra in eighth grade has unwanted consequences. His study of New Hampshire schools showed that "only about half of the students who take algebra as eighth graders continue their study of mathematics through a fifth year." Prevost recommends an enriched program rather than algebra, with only the top 3 to 5 percent being truly accelerated. Two very relevant questions regarding his study are:

1. Do the sixth and seventh grade programs adequately prepare students for algebra in eighth grade?
2. Would the retention rate be greater if there were alternative twelfth grade courses?

Several letters of rebuttal were submitted by readers in response to Prevost's article.

There has been specific concern about students who study calculus in grade 11. ETS reports that 4,000 of the 60,000 students now taking AP calculus exams are in eleventh grade and lower. Very special arrangements are needed to guarantee that these students have appropriate mathematics in grade 12.

The NCTM's position statement on Provisions for Mathematically Talented and Gifted Students (October 1986) contends that:

> The needs of mathematically talented and gifted students cannot be met by programs of study that only accelerate these students through the standard school curriculum, nor can they be met by programs that al-

low students to terminate their study of mathematics before their graduation from high school.

The NCTM paper goes on to recommend that:

All mathematically talented and gifted students should be enrolled in a program that provides a broad and enriched view of mathematics in a context of higher expectation. Acceleration within such a program is recommended only for those students whose interests, attitudes, and participation clearly reflect the ability to persevere and excel throughout the entire program.

Greater Integration of Topics

It can hardly be disputed that the curriculum for grades 11-13 must be closely related to the curriculum of the preceding years. An issue on which there is no apparent consensus and which therefore requires further study is the extent to which the secondary school mathematics curriculum should be integrated or unified.

In most countries mathematics is not compartmentalized into algebra, geometry, etc., as is conventional in the U.S. Since World War II there have been numerous attempts to break down these compartments in this country. One such attempt in the 1950's was Florida's Functional Mathematics Program which was short-lived mainly because of the advent of the School Mathematics Study Group (SMSG) and the "new mathematics." In the late 1960's and early 1970's, the Secondary School Mathematics Curriculum Improvement Study (SSM-CIS), directed by Howard Fehr at Columbia, developed a unified program intended for the top 15 to 20 percent of secondary school students.

In 1984, after several years of experimentation, the New York State Board of Regents adopted an integrated approach as the high school Regents program in mathematics with intentions of gradually phasing out the traditional program. At the present time, the University of Chicago School Mathematics Project is developing an applications-oriented curriculum that is integrated to some extent although algebra prevails in grade eight and geometry receives the emphasis in grade nine.

Despite these moves toward unification, it is still the case in the U.S. that most college preparatory mathematics programs begin with a year of algebra, followed by a year of geometry and another year of algebra. The Second International Mathematics Study Summary Report for the United States (1985) contends:

It is plausible that the "fragmentation" and "low intensity" found in many of our mathematics programs could be allayed by a more integrated approach to the high school mathematics curriculum.

The Report of the MSEB Task Force on Curriculum Frameworks for K-12 Mathematics (Draft, October 1986) states:

Not only do we believe that an integrated curriculum offers the possibility of a richer, more coherent program than the alternative but, further, we believe that the introduction of a variety of new subject matter into the secondary school mathematics curriculum will inevitably signal the demise of the segregated curriculum if only for logistics reasons.

How inevitable this is remains to be seen. One factor that must always be taken into consideration when making curriculum decisions in the U.S. is the mobility of our population. It must be admitted that the ability to transfer a credit in algebra or a credit in geometry has simplified matters for many students. Unified curricula may make the process of transferring more traumatic. Such curricula may also cause much rethinking on the part of colleges that are accustomed to accepting well defined units of credit, although placement by examination may suffice. An additional concern is that teaching unified courses requires a breadth of understanding beyond what many teachers have been prepared to provide.

It should be pointed out that there is yet another aspect of integration that needs further study. This is the possibility of integrating mathematics with other disciplines, particularly science.

Role of Statistics & Discrete Mathematics

There appears to be agreement that topics in discrete mathematics and in statistics and probability should be included in the mathematics curriculum for college-bound students. However, there is lack of agreement on the appropriate place for these subjects in the mathematics curriculum and on the number of hours of study required, especially for students who study calculus in grades 12 or 13. The traditional mathematics curriculum from elementary algebra, geometry, intermediate algebra, through precalculus is largely a calculus preparatory curriculum. Usually these courses do not include substantial study of discrete mathematics or of statistics and probability even though foundational topics in these areas are often in the back of textbooks used in the courses.

STATISTICS AND PROBABILITY

Nearly every major committee making recommendations on the high school curriculum has said that familiarity with the basic concepts of statistics and statistical reasoning should be a fundamental goal for high school mathematics (new state frameworks in California, Illinois, Wisconsin, and New York; The College Board's *Academic Preparation in Mathematics;* MSEB Task Force Draft Report on Curriculum Frameworks for K-12 Mathematics; *A Nation at Risk* from the National Commission on Excellence in Education). Typical is the statement in *Educating Americans for the 21st Century* (1983) from the National Science Board Commission on Precollege Education in Mathematics, Science and Technology: "Elementary statistics and probability should now be considered fundamental for all high school students." Similarly, after studying the performance of 12th grade college preparatory mathematics students on the Second International Mathematics Study, the U.S. National Committee recommends that, "The curriculum should be broadened and enriched by including a substantial treatment of topics such as geometry, probability, statistics and algebra, as well as promoting higher-level process goals such as estimation and problem-solving."

Committees have been making recommendations for the inclusion of statistics for thirty years. In 1959, the Commission on Mathematics of the CEEB in its report, *Program for College Preparatory Mathematics,* recommended a one-semester course in probability and statistics as an alternative for grade twelve. The Commission published an experimental text, *Introductory Probability and Statistical Inference,* the same year. In 1975, the Euclid conference sponsored by the NIE identified probability and organization and interpretation of numerical data as two of ten basic goals for mathematics education. Also in 1975, the Conference Board of the Mathematical Sciences National Advisory Committee on Mathematical Education (NACOME) reported:

> While probability instruction seems to have made some progress, statistics instruction has yet to get off the ground . . . .

The situation has not changed much since this NACOME report. For example, Bruce Williamson in *The Statistics Teacher Network Newsletter* (1983) reported that a study of approximately 350 high schools in Wisconsin found that the percentage of schools which allot more than three weeks in the total high school program to statistics declined from 26% in 1975 to 23% in 1983. In 1975, 43% allotted more than three weeks to probability; this declined to 34% in 1983. However, Wisconsin now requires some elements of statistics for all students so it is likely the decline has been reversed.

The dearth of good materials may be the main reason why more statistics and probability is not being taught. Usiskin (1985 NCTM Yearbook) states, "The content of any new curriculum must be specified in as much detail as current content. . . . Materials must be available to implement recommendations." Materials currently available consist primarily of locally produced handouts, a recently published book for high school students (Travers, et al.), and four booklets published by the ASA-NCTM Joint Committee on the Curriculum in Statistics and Probability.

Arguments against including more statistics are raised by educators who feel that students in grades 11-13 need to spend their time on "basic" mathematics topics. For example, the integrated New York State mathematics program is criticized by the IEEE Long Island Section because the introduction of new subjects such as statistics and probability reduces the time spent on basic algebra, geometry, and trigonometry.

DISCRETE MATHEMATICS

Recently, topics in discrete mathematics have been recommended for inclusion in the high school curriculum in several reports including the Report of The MSEB Task Force on Curriculum Frameworks for K-12 Mathematics (Draft, October 1986). The CBMS ("The Mathematical Sciences Curriculum K-12: What Is Still Fundamental and What Is Not") recommends that discrete mathematics now be regarded as "fundamental."

While certain topics (induction, matrices, discrete probability, and combinatorics) are found at the back of many high school textbooks, they are not always taught. Additional topics such as graph theory, difference equations, recurrence relations, and game theory are also recommended by some. No high school curriculum has yet been standardized in discrete mathematics. Some relatively short units, such as the HiMap modules, are now available and finite mathematics texts are sometimes adapted for this instruction.

Many colleges and universities now offer a lower division course in discrete mathematics particularly suited to students in computer science. There has been discussion of how students should be prepared for such a course. To quote from the preliminary report (1984) of the MAA Panel on Discrete Mathematics in the First Two Years:

> What should be taught in the high schools or on the remedial level in the colleges to prepare students adequately for this course? Our suggestion is tentative: some of us feel that perhaps a revived emphasis on the use of both formal and informal proof in geometry courses as a means for teaching methods of proof and analytic thinking would be a step in the right direction. Others of us are not so sure. Increased use of algorithmic thinking in problem solving could be easily adapted to many high school courses . . . . Simple restoration of some of the classical topics (the binomial theorem, mathematical induction, natural logarithms) and increased emphasis on problem solving might make the proposed course much easier for the student.

The issue of what curriculum adequately prepares students for college-level discrete mathematics or whether any particular preparation is essential requires further study.

Twelfth Grade Mathematics Courses

For students in a college preparatory program who take algebra I in grade 9, the twelfth grade course is traditionally a year of precalculus mathematics that includes trigonometry as well as topics such as exponential and logarithmic functions and equations, conic sections, rational functions and their graphs, polar coordinates, parametric equations and their graphs.

Two groups of students are identified by some as students who need alternatives to traditional twelfth grade courses. The first are students who have studied precalculus in grade 11 but who may not benefit particularly from the study of calculus in grade 12 and who may benefit more from the study of other topics in mathematics, delaying calculus until grade 13. The second are students who do not expect to need calculus in their college programs, but do expect to take college mathematics. It is argued by some that standard precalculus in grade 12 (or calculus, in the case of students who are eligible for calculus) does not provide these students with the best college preparation.

The three most frequently mentioned semester-length courses proposed as 12th grade options are given below. We also include a new course under development at the North Carolina School of Science and Mathematics.

STATISTICS AND PROBABILITY

A one-semester course in statistics and probability has been proposed by the new *Mathematics Framework for California Public Schools*. The tenth grade course being designed by the University of Chicago School Mathematics Project is Statistics and Computers. Currently, statistics courses are not widely taught. For example, of the 42 high schools in New Hampshire, only five offer a course in statistics (Prevost in *The Statistics Teacher Network Newsletter*, 1983).

DISCRETE MATHEMATICS

Using their own lecture notes, Georgetown University lecturers have taught a summer course in discrete mathematics/mathematical modeling to selected high school students (Sandefur in 1985 NCTM Yearbook). The North Carolina School of Science and Mathematics has offered a one-semester course called "Topics in Discrete Mathematics" which follows multivariate calculus, again to very select students. We know of no experiments in traditional high schools. The curricula

from the six colleges and universities that were funded by the Sloan Foundation to integrate discrete mathematics into the first two years of the college program may provide some guidance for secondary schools seeking a 12th grade course in discrete mathematics.

LINEAR ALGEBRA

Full courses in linear algebra are not common in secondary schools although various individuals argue the appropriateness of this mathematics for grade 12 (e.g., John Thorpe, "Algebra: What Should We Teach and How Should We Teach It," NCTM Research Agenda Project Conference on the Teaching and Learning of Algebra, Athens, Georgia, March 25-28, 1987). In the Chicago area, high school students involved in accelerated programs through the Johns Hopkins talent search do take linear algebra in several junior colleges.

A SURVEY OF MODERN MATHEMATICS

A twelfth grade course is being developed with a Carnegie Foundation grant at the North Carolina High School of Science and Mathematics that is a survey of modern mathematics. The year-long course will consist of units (at least three weeks long) which introduce students to the kinds of mathematics that they could study in college. Students will presumably then be able to make a more informed choice of their first college courses. The topics proposed include calculus, discrete mathematics, computer programming, popular software such as SMPs, statistics, probability, mathematics of finance, linear programming, operations research, and linear algebra. The course focuses on the types of problems that are characteristic of each field and introduces students to the mathematical techniques that are used to solve them.

Calculus Review for College Freshmen

Each year a large number of students take a full year of high school calculus and either do not write the AP test or do not receive a 3, 4, or 5 on it, and do not test above the beginning calculus level on college placement tests. These students may not "fit" into a standard college calculus sequence. They typically view the introductory material as mathematics they have already learned and do not take beginning work seriously enough to succeed in the later work of their courses. It can be argued that these students are potentially capable students in mathematics. The CUPM Subcommittee on Calculus Articulation [Don Small, Chair] has recommended that colleges should develop a special course with the following characteristics:

1. The course should be different from high school calculus.

2. The course should contain a broad review component designed to provide depth missing in most high school courses.
3. The course should assume a high school calculus experience and build on it.
4. The course, when completed successfully, should provide one semester of beginning calculus credit in addition to the credit for the course.

Colby College has developed a two-semester calculus sequence integrating the treatment of one and several variables that has these characteristics. The course is offered for identified students instead of the regular three-semester calculus course.

On the other side, many argue that if colleges develop special courses for students who take calculus in high school but do not master it at a college level, students will be encouraged to be satisfied with less than full mastery of their high school calculus and high schools will be encouraged to offer watered-down courses in calculus in the 12th grade. (The MAA and the NCTM have prepared a joint statement for the schools indicating that students who take high school calculus should expect to establish college credit either through the AP exam or through a college proficiency exam.) These educators maintain that alternatives to calculus should be developed in grade 12 for students who are not ready to master calculus, and that it is not a good use of student time to spend one and a half to two years on the content of first-year calculus.

The Place of Deductive Reasoning

University and college faculty complain that too often students enter college with the view that mathematics is just a collection of rules and algorithms to be used to attack a variety of standard problems. They assert that students' abilities to reason either deductively or inductively have not been developed by their mathematics courses, and that their ability to attack problems of an unfamiliar nature has not been developed. If these statements are true, there is need to determine the cause. Some argue that the rush to streamline the high school program so that calculus can be taught in the 12th grade, jettisoning things deemed unimportant to this goal and omitting end-of-the-text topics, bears primary responsibility. Others claim that the emphasis on the formal structure of mathematics has stifled the ability to reason intuitively. Also it is not clear how much better entering college students a generation ago were at mathematical reasoning. There is general agreement that strengthening students' reasoning ability is a goal in grades 11-13, but little evidence that this is happening. The following comments by Phil Curtis, UCLA,

suggest where attention may need to be focused:

If this problem is to be corrected, more attention must be paid to the development of a student's reasoning ability at all levels, and the instruction should focus on inquiry and discovery rather than an over-riding emphasis on mastery of a body of rules and techniques. Is the following statement true or false? How can we decide? If true, why is it true? How can we see if the statement is a consequence of ideas that have been discussed before?

The emphasis on reasoning should focus on intuitive understanding rather than a development of formal logic. The structure of implication, however, should be stressed. What is the nature of mathematical implications? What is the hypothesis and what is the conclusion of an implication? What is the difference between a statement and its converse?

Traditionally, the geometry course was the first course in which students met the notion of mathematical proof. Proofs were very formal but, since there was never an attempt to extend this form of reasoning to other mathematics courses, very little in the way of reasoning skills was retained when students came to college. There are many areas where mathematical implication can be stressed: solution of equations and inequalities in algebra as well as certain elementary ideas from number theory, use of coordinate geometry and vector techniques in geometry, mathematical induction, and counting arguments and elementary probability in advanced algebra or discrete mathematics.

To develop mathematical reasoning skills takes time and should be one of the primary goals of the high school program. Ideas should be developed leisurely; the focus on the program should not be just to make students ready for calculus in the 12th grade. To accomplish this development the program needs to be opened up at all levels. In particular, the necessity for inordinate amount of review present at all levels should be lessened. Rule bound students are on a mathematical dead end. The high school program should be able to do more.

Curricular Impact of Calculators

The introduction of single variable calculus as the "desired" 12th grade mathematics course in high school is thought by some to have resulted in a streamlining of the program prior to calculus. The resulting lack of seasoning or maturity on the part of many incoming college freshmen is perhaps the biggest complaint of college faculties. Students are said to know much more than students of a generation ago as far as calculus-related ideas are concerned, but geometric intuition, skill with coordinate geometry, the ability to organize applied problems and the ability to construct mathematical arguments are all too often reported to be lacking. Students may have an algorithmic facility with mathematics, but many lack a true understanding of the subject.

It is often expressed that when hand-held calculators are generally available at all grade levels the narrowing of the curriculum may get worse. This apprehension is typified by the remark: "If students are to use calculators, why should we teach all of this arithmetic (and possibly algebra) that can be done so much easier on a calculator?" Experience in other countries, e.g., Australia, shows that narrowing of the curriculum need not be a serious problem.

Indeed, many argue that an introduction of calculators can encourage a broadening of the curriculum rather than a further narrowing of it. At the elementary level, students can handle numerical data in much greater amounts and in wider practical situations than would be possible without calculators. A sense of 'reasonableness of solution' to more complicated calculational problems can and should be a goal of instruction. There is an increased opportunity to develop a student's mental arithmetic and estimation skills.

Rather than decrease students' algebraic skills, these proponents argue, the ability to confront practical problems of much greater computational complexity should be an impetus for development of the algebra skills necessary to organize the problems so they are amenable to numerical calculation. All of the calculations involving compound interest are possible, with the associated opportunity to do manipulations with geometric series. Transcendental equations in trigonometry are easily solved. This could be a spur to confront more meaningful applied problems in trigonometry involving manipulation and solution of trigonometric equations; computations that were impossible in the past. Certain topics of course can be dropped; for example, dependence on tables for teaching trigonometry as well as the use of logarithms to perform multiplication and division calculations.

Curricular materials are under development now on several levels and at several locations that make central use of calculators and computers. The answers to many questions concerning the effect of technology on the curriculum will emerge as these materials become widely available. For grades 11-13 the content of intermediate algebra, precalculus, discrete mathematics, and calculus are all likely to be affected. Ronald Douglas has noted in the Introduction to *Toward a Lean and Lively Calculus* (MAA, 1987):

> Anyone who has seen hand-held calculators which output the graph of an equation usually realizes that we can and, indeed, we will have to change what we ask students to learn and what we test them on. And this is just a start of developments that include the growing availability of programs for symbolic and algebraic methods as well as for numerical methods.

The Sloan Foundation has funded projects at seven colleges and universities to consider the potential impact of computer algebra systems on the teaching of calculus. The National Science Foundation has announced a major initiative in the area of calculus curriculum that can be expected to include some projects where calculators and computers play a central role.

Alternatives to Remedial Courses

An ever increasing fraction of resources is now devoted to remedial courses at both the college and high school levels. Algebra and precalculus courses are common at colleges for students who previously have been exposed to this material. Failure rates of 40-60% are reported in beginning algebra courses at the 10th grade level. At both levels, part of the problem seems to be that the necessary prerequisite material was not learned; and the remedy often is to teach material over again in the same way it was taught the first time. Some argue that there are alternatives.

Initially, students who have no prospect of success should not be placed in a beginning algebra course. There are two problems here: the need for an effective predictor test and the need for middle grade curricula to better prepare students for algebra. The California Mathematics Diagnostic Testing Project is developing an algebra readiness test. Several curriculum projects for the middle grades are seeking to strengthen curriculum at that level. A common assumption in these projects is that imaginative use of hand-held calculators, with the tremendous increase in calculational power they give students, should play a central role in stimulating student interest and providing different perspectives on the abstract ideas, for example that of a variable, that students will encounter.

There are proposals that more flexible scheduling could also ease the remediation problem. These proposals argue that students who fail the first semester of beginning algebra should have the opportunity of repeating it in the second semester and not be forced to wait until the following year to begin again. However, this flexibility is seldom present in school schedules. In some districts, students who fail an entire year are given the opportunity of making the course up in a summer session. When a student attempts to learn a year of algebra I, geometry, or algebra II in a shortened summer session, there is little chance the student will become proficient with the fundamental ideas of the course.

Some geometry texts attempt to help avoid remediation, or excessive review, at the intermediate algebra level by providing diagnosis in the geometry course of what algebra skills have been retained and what have not. A renewed emphasis on these topics parallel to the development of geometry can be used to achieve readiness for the second course in algebra. Coordinate geometry is a possible framework for renewal work in algebra.

In some colleges and universities, algebra and precalculus courses avoid being just a review of high school material by making actual use of computers and calculators and by blending precalculus topics with an introduction to the notion of limit and other fundamental ideas from calculus.

Use of Standardized Test Scores

Standardized tests are widely used in high school and beginning college mathematics programs throughout the United States. These tests range from achievement examinations, such as the mathematics achievement tests of the College Board and the New York State Regents Examinations, to mathematics aptitude tests such as the SAT, and various assessment instruments designed at the state level and often used at the 12th grade and other levels. In addition there are diagnostic instruments designed to assess readiness for the next level of the program and placement examinations, such as those available from the MAA, often used at the college freshman level and in high school.

These tests are often criticized on various grounds. First, since they are usually multiple choice exams, they are criticized for supposedly not testing higher order thinking or problem solving skills. Secondly, since the general assessment exams are used to compare schools and programs, there can be considerable political pressure "to get those scores up," with the result that there can be a narrow concentration by the teacher on just those basic skills covered in the examination.

Also, tests designed for one purpose are often used for another. Diagnostic examinations are used as assessment instruments to compare the performance of classes and schools. When this is done the pressure to narrow the curriculum to just those basic skills necessary for success in the next course can become great. On the other hand if this comparative pressure is removed, diagnostic tests can be quite effective in indicating areas of the curriculum which are not being retained by the student but which are absolutely necessary for success at the next level of the program. Strategies for dealing with these deficiencies can then be constructed which should result in a broadening of the curriculum rather than a narrowing of it to a concentration on just basic skills.

Placement exams can be misused if they are used as a barrier which prevents students from taking a given mathematics course, rather than providing a route covering necessary preparatory material. The latter then places the student in the course more properly prepared.

Other misuses of tests are commonly cited: standardized tests designed for comparative assessment purposes that "drive the curriculum," failure to allow for margin of error in interpreting scores, excessive use of multiple choice tests for classroom assessment.

The Mathematical Sciences Education Board has designed a comprehensive testing study that will survey testing practices, assess how test results are used, and the effects of testing on curriculum and teaching behavior. This study should provide a foundation for future decisions related to test construction and use.

The Form Geometry Should Take

There is general agreement that many entering college students lack geometric intuition and the ability to visualize geometric situations. This would seem to indicate that a strengthening of the geometric content of the high school program is sorely needed. But there is not a consensus as to how this should be done. Nor is there agreement on the place of such topics as transformations and vectors or the amount of emphasis that should be placed on formal proof or the desirability of introducing students to formal logic within the geometry course.

What ought to be taught in high school geometry has been debated for at least a half century. In an address to the 1958 annual meeting of the NCTM in Cleveland (published in the *Mathematics Teacher* as "The Nature and Content of Geometry in the High Schools"), Julius Hlavaty referred to the "continuing crisis in the teaching of geometry" that had endured for "fully 25 years." Titles of other articles in the *Mathematics Teacher* in the last two decades substantiate the existence of an ongoing debate (Adler, "What Shall We Teach in High School Geometry?" March 1968; Allendoerfer, "The Dilemma in Geometry," March 1969; Fehr, Eccles, and Meserve, "The Forum: What Should Become of the High School Geometry Course?" February 1972). The introduction to the 1973 NCTM Yearbook *Geometry in the Mathematics Curriculum* is entitled "Disparities in Viewing Geometry." The book contains chapters discussing several different approaches to high school geometry: conventional, coordinate, transformation, affine, vector, and an eclectic approach. In the latest yearbook of the NCTM, *Learning and Teaching Geometry K-12* (1987), Usiskin writes on "Resolving the Continuing Dilemmas in School Geometry," and Niven on "Can Geometry Survive in the Secondary Curriculum?" Usiskin offers suggestions for resolving the dilemmas, and Niven proposes recommendations that he thinks will make geometry a more attractive subject. Other chapters in the Yearbook discuss various geometric topics and applications.

The proliferation of computers, the growing popularity of Logo, and the development of software such as "Geometric Supposer" are adding new elements to the debate on geometry. Can they perhaps help to resolve issues that have been around for quite some time?

Collaborative Efforts

In recent years there have been many studies and reports released calling for significant changes and reforms in American education. While there has been considerable consensus in the identification of the ills within this system, there has not been universal agreement as to what the solutions should be in order to remediate the identified problems and shortcomings. Such has been the situation with issues involving education in mathematics, particularly those related to curriculum.

Issues related to curriculum have occupied an important position in many of the released studies and reports dealing with education in mathematics in the United States. In particular, discussions and recommendations pertaining to the mathematics curriculum of grades 11-13 have been prominent in many of the published reports and studies. Some of the consensus and lack of consensus items pertaining to the mathematics curriculum have been identified and dealt with in other portions of this report. While a listing of these items is the major thrust in this report, it is necessary to identify briefly several other considerations which pertain to the total mathematics experience for 11-13 grade students.

Many of the recent reports have cited the success of mathematics students in other countries, particularly those from Japan. The reports emphasize that the success of the students is due in part to the collaborative efforts of the school and family, and due to the emphasis, status, and importance accorded to learning and education. Enhanced success in mathematics education is dependent on extensive collaborative efforts involving several subgroups. We have recommended that targeted brochures and flyers be developed and circulated to these groups. We wish here to discuss further the various roles these groups must play in the desired collaboration and also the essential role of professional organizations in creating a climate for change.

Extensive partnership and collaborative efforts should involve school leadership, primarily the principal; public guidance personnel, primarily guidance counselors; the home, primarily the parents; the mathematics professional community, primarily the classroom teachers; and of course, the students. Collaboration within the schools needs to be linked also to colleges and universities. In these institutions, in addition to mathematics faculty members, admissions personnel play key roles. Some prototype regional programs for collaboration include the Bay Area Mathematics Project, and the Mathematics and Science Education Network of North Carolina. The more recently formed American Mathematics Project is chaired by R.O. Wells, Jr., of Rice University. It aims at encouraging and extending local cooperative efforts involving elementary, junior high school, and high school teachers, college and university faculty, and professionals in industries.

In the paragraphs which follow, several issues and questions relative to each of the essential groups will be addressed to provide evidence for a need of such partnerships. Through the raising and addressing of such issues it is apparent that educators and professional organizations will need to take advantage of many opportunities in collaborative efforts in order to affect curricular changes. It is also evident that curricular issues in and of themselves cannot be considered apart from factors which will have a significant effect upon any curricular proposals or modifications.

There are numerous issues related to the leadership within the schools today, and there is research related to the role of leadership, particularly the role of principals, within successful learning environments. Administrators, as agents of Boards of Education, need to become partners with mathematics educators in dealing with curriculum-impacting issues such as the following:

1. The recruitment and retention of qualified staff members.
2. The development and maintenance of educational settings and environments which promote quality instruction, effective learning, and the maintenance of academic standards.
3. The providing of support and encouragement for professional growth, and the upgrading of content competencies and professional teaching skills.
4. The providing of support which enables the opportunities within schools for curricular change.
5. The providing of settings which enable participation in leadership by those having expertise related to education in mathematics.
6. The involvement of administrators as advocates for appropriate changes in the mathematics curriculum.

A second area where collaborative efforts need to be fostered involves the school guidance personnel. Efforts are needed to provide aid and support to counselors who have significant impact upon students. (In March, 1987 the NCTM made several recommendations concerning the advising of students in mathematics in the statement, "Counseling Students in Planning Their Mathematics Programs.") Often, guidance counselors do not have, nor can they be expected to have, a comprehensive understanding of the content or importance of the 11-13 grade mathematics curriculum. Mathematics teachers and guidance counselors need to address the following issues together:

1. How should students be placed in appropriate courses?
2. Which mathematics courses do students need in preparation for other mathematics courses, careers, and future academic work?
3. What communication with students and parents is necessary in order to direct students into appropriate mathematics courses?
4. What support should counselors be able to provide to encourage students in their studies of mathematics?
5. What support should counselors be able to provide to students who are struggling in their mathematics studies?
6. What efforts are needed to encourage all groups of students, regardless of gender or race, to take more mathematics courses?

Many of the recently released reports attribute the success of students in mathematics to the attitudes and the values developed in the home. Parental influence has been shown to have a significant impact upon the successes of the child in school. There are numerous issues and questions which need be addressed in developing stronger partnerships involving parents. (In Spring 1987 the National PTA sent a special mailing to the 25,000 local PTA presidents describing comparative data from international studies in mathematics and calling on parents to be more involved in finding solutions to problems in mathematics education in the schools.) Items for consideration include the following:

1. Parents need to provide an appropriate environment which encourages home study and the completion of homework.
2. Educators and parents need to work together to establish attitudes and priorities where learning becomes more valued.
3. There needs to be a support system within which class attendance is a high priority.
4. Parents need to become better informed about the need for appropriate mathematics backgrounds for their children. It is not enough that students take mathematics courses. They should take mathematics courses appropriate for their career goals.

5. Parents themselves need to be educated about the learning of mathematics. To many adults the study of mathematics is thought to be the performance of arithmetic computations. It needs to be stressed that the learning of mathematics is a continuous, sequential process to be pursued over a period of time. Parents need to understand the need for regular practice in the learning of mathematics.

Another component in the collaborative partnership is the teacher. Until the teacher becomes convinced that there is a need to improve the 11-13 grade curriculum, advances in the quality of 11-13 grade mathematics programs will be limited. Professional organizations will have significant opportunities to provide leadership, encouragement, and training to teachers. Significant thought and attention will be required to bring teachers into the process of affecting change in mathematics.

Students need to be involved fully in the educational process. They need to be made aware that an appropriate mathematics background is a necessity and that mathematics is a "hands-on activity" and a "do-it-yourself" activity. The teacher can be an aid or facilitator of learning, but cannot do the learning for the students. In the era of instant gratification it is important that students realize that solving a mathematical problem will not always be quick nor will it be easy; in fact, not every problem will have a solution. Students need to develop persistence and to realize that progress will often seem slow, especially in the reading of mathematics. Students also need to understand that it is important to keep up with their assignments and that class attendance is essential for success in the study of mathematics.

In the preceding paragraphs several subgroups have been identified as being necessary components for effective partnerships in addressing the issues and questions associated with the 11-13 grade mathematics curriculum and instructional program. The primary responsibility of this task force study has been to focus upon the areas of consensus or lack of consensus as identified in prior reports and studies. Nevertheless, it seems appropriate to address some of those issues which ultimately will determine much of the success or failure related to those curricular items for which there is consensus. There needs to be collaboration of school leaders, counselors, parents, teachers, and students—and linking with college and university personnel—in order to improve mathematics programs for the 11-13 grade student.

This Task Force has recommended that the NCTM and MAA prepare brochures and flyers targeted specifically at the various groups described in this section.

There are additional roles that the professional organizations will need to play if the collaborative efforts of these groups are to succeed. We urge the organizations to continue to enable greater communication among the various groups in these ways:

- To increase and expand the design of workshops and programs for policy makers, high school principals, and post high school level administrators in order to address effectively the issues relating to the recruitment and retention of qualified instructors and teachers of mathematics.
- To expand efforts to work at national, state, and local levels with and through the affiliated groups on issues related to the recruitment and retention of qualified mathematics teachers and instructors.
- To expand their roles in working with affiliated and non-affiliated groups in providing forums to address such issues as curriculum change, uniform standards of quality instruction, equity of opportunity, and expectations in mathematics education.
- To focus further on ways of being more effective in communicating the solutions to identified problems beyond their membership in order to provide the appropriate and necessary impact upon policy makers, administrators, counselors, teachers, parents, and students.
- To undertake actions and assume a significant role in developing programs where articulation with guidance counselors becomes a major priority.
- To work with school administrators, counselors, and teachers in establishing programs which will enhance and increase parental involvement and support in the educational process.
- To develop strategies that will involve the students more deeply in the educational process.

Task Force Members

Executive Committee

JOAN R. LEITZEL, CHAIR, The Ohio State University.

PHILIP C. CURTIS, University of California, Los Angeles.

CHARLES HAMBERG, Illinois Mathematics and Science Academy.

DONOVAN R. LICHTENBERG, University of South Florida.

ANN WATKINS, Los Angeles Pierce College.

Reading Committee

BETTYE ANNE CASE, Florida State University.

DON CHAMBERS, Wisconsin State Department of Education.

DWIGHT COBLENTZ, San Diego, California.

STEVE CONRAD, Roslyn High School, New York.

RONALD G. DOUGLAS, State University of New York, Stony Brook.

WADE ELLIS, JR., West Valley College.

DALE EWEN, Parkland College.

JEROME GOLDSTEIN, Tulane University.

KATHLEEN HEID, Pennsylvania State University.

DAVID JOHNSON, Nicolet High School, Wisconsin.

KATHERINE P. LAYTON, Beverly Hills High School.

PETER LINDSTROM, North Lake College.

WILLIAM LUCAS, Claremont Graduate School.

IVAN NIVEN, University of Oregon.

DONALD SMALL, Colby College.

THOMAS TUCKER, Colgate University.

Minimal Mathematical Competencies for College Graduates

This chapter contains the report of the CUPM Panel on "Minimal Mathematical Competencies for College Graduates," reprinted from the AMERICAN MATHEMATICAL MONTHLY, *89 (April 1982) 266-272. Donald Bushaw, chair of the panel, has prepared a new preface relating issues addressed by the panel to many themes that are part of today's debates about higher education.*

1989 Preface

On Thursday, December 15, 1977, the Carnegie Foundation for the Advancement of Teaching released its famous report "Missions of the College Curriculum." This report, which received a great deal of attention at the time, described general education in U.S. colleges and universities as "a disaster area," and expressed special concern about the neglect of mathematics and English composition.

The following Monday, Henry L. Alder, then President of the MAA, wrote a letter challenging the MAA's Committee on the Undergraduate Program in Mathematics (CUPM) to take up the matter, suggesting as one possibility the formation of a "new CUPM panel or subcommittee" to "consider the problem of general education in mathematics for all or most college students."

At the CUPM meeting of January 8, 1978, Chairman William F. Lucas appointed a subcommittee ("panel" in the then current nomenclature) to do just that. After a considerable amount of study and discussion, and several diverse surveys, the panel presented its brief and temperate report to CUPM, which approved it.

The continuing turbulence surrounding the idea of general education—witness the unexpected popularity of the recent books by Bloom and Hirsch—is evidence that not all of the problems set forth in the 1977 Carnegie Report have been solved. Many colleges and universities have made, or are still making, major revisions of their general education programs, and mathematics (often under the guise of "quantitative thinking," "computation," or the like) is a frequent theme in the concomitant discussions.

Within this setting, CUPM's 1982 report seems to stand up well. If it were to reconvene today, the panel would certainly reaffirm all of its recommendations, and none more strongly perhaps than Recommendation E, which presents an eminently sensible and even exciting idea that seems to have been carried into action in very few places.

One would like to think that need for the "remedial" course sketched in the report has declined, or will soon decline, because of nation-wide attention to weaknesses in the precollege mathematics curriculum. In any case, the course itself should still be useful for whatever remains of the clientele for which it was intended.

Courses in mathematics appreciation meeting the standards implied in the report are probably still rare, although courses of similar intent are not uncommon.

If a survey of persons from the Combined Membership List were redone today, the responses might show more interest in discrete mathematics, and might show effects of the rapid progress in the design and dissemination of calculators and microcomputers in the intervening years; but the responses given almost ten years ago tended to be conservative, and a new round of responses would probably tend to be conservative too.

Thus the report, though neither radical nor voluminous, presents some worthwhile ideas that are still far from commonplace, and which, if widely adopted, could contribute significantly to the mathematical competence and maturity of coming generations.

Donald W. Bushaw
Washington State University
March, 1989

Introduction

Too many people know too little mathematics. Even those who are well informed in other ways often cannot appreciate, much less participate in, some major currents of modern life because of their ideas and feelings about mathematics. In a relatively severe but all too common form, ignorance of mathematics amounts to a form of "functional illiteracy."

Along with the recent revival of interest in general education, "core" curricula, and minimal competencies, this problem has naturally led to the question: What mathematics should every graduate of an American college or university know?

At its January 1978 meeting, the Association's Committee on the Undergraduate Program in Mathematics (CUPM) established a panel to study the question and make appropriate recommendations. Some of the work of the panel is described in an Appendix to this document, which is a report from the panel.

The recommendations and other ideas set forth in this report will surely not be the last word on the subject. Many intelligent people will be giving further thought to it, and future experience should certainly be allowed and expected to affect our outlook on the whole matter.

Recommendations

The leading lesson the panel learned from its surveys (see the Appendix) is that American colleges and universities are so diverse that it is impossible to describe either an approximately standard practice or an everywhere attainable goal. A set of minimal competencies that might be woefully inadequate for specialized or selective universities can be a hopeless ideal for others. To perform its task realistically, the panel has therefore felt obliged to interpret the word "minimal" in a really minimal way. *The recommendations listed below accordingly refer to a bare minimum of mathematical competencies for all college graduates.* The panel hopes that individual institutions will go as far beyond these recommendations as local conditions allow. Similarly, how the requirements should be met is left open, for that depends not only on the requirements themselves but also on local policies, traditions, and resources.

The following recommendations result from the panel's studies and deliberations. In preliminary form, they have been reviewed by numerous mathematicians and nonmathematicians, and have been considerably modified in light of comments received. In this sense they represent the collective judgment of a group much larger than the panel itself.

RECOMMENDATION A:

All college graduates, with rare exceptions, should be expected to have demonstrated reasonable proficiency in the mathematical sciences. Every college or university should therefore formulate, with adequate concreteness, what this "reasonable proficiency" should mean for its students; define how students should demonstrate this proficiency; and establish this demonstration as a degree requirement.

Competence in arithmetic and some facility in making applications in everyday life might be a reasonable graduation requirement for two-year college students in terminal and vocational programs.

Four-year colleges and universities should normally require—perhaps on entrance—not only these but elementary algebra and elementary geometry. They

should also expect graduates to understand and be able to use some elementary statistical ideas, to be aware of the place of mathematics in society generally, and to appreciate the nature and societal significance of computing. This applies also to two-year college students in university parallel curricula.

RECOMMENDATION B:

Whether or not stipulated proficiency is tested by examination, courses should be made available in which it may be acquired. These courses should be taught by effective instructors, and should be designed to be appealing and significant to the students.

RECOMMENDATION C:

In particular, one or more courses of a remedial nature should be available where there is a need. Such courses, by definition, ordinarily present precollege material, but it should be presented in a way suited to the clientele. In institutions where it is considered improper or impossible to offer remedial courses, mastery of the mathematics should be assured either by entrance requirements or by referring students to other schools where remedial courses can be taken. Two-year colleges have made a large contribution in this role and may be expected to continue to do so.

Is college credit appropriate for remedial courses? On this point we will only quote the statement approved by the MAA Board of Governors on August 20, 1979: "College credit granted for work in mathematics must be carefully controlled. It should not be granted for distinctly high school level work. Mathematics courses offered in college should be examined to determine the extent of their overlap with high school mathematics, and where that overlap is substantial the course should not provide credit toward college graduation; but the students should be graded on their work, and the results should be included in computing grade point averages."

RECOMMENDATION D:

While almost all undergraduate courses in mathematics should give attention to applications and to historical and philosophical aspects of the subject, there should be one or more courses that concentrate on these aspects while remaining accessible to students with little mathematical background.

RECOMMENDATION E:

Individual interests often lead students to take a considerable amount of post-secondary mathematics in conventional courses. These students should also be able to take a course of the kind described in Recommendation D, but presupposing more mathematical background.

The MAA Committee on Improving Remediation Efforts in the Colleges, chaired by Professor Joan Leitzel, has gathered information about effective remedial programs and has made its own recommendations. A separate CUPM panel, chaired by Professor Jerome Goldstein, is at the same time formulating recommendations on "mathematics appreciation" courses of the kind described in Recommendations D and E and in the second section below. (The full report of this panel is reprinted in the following chapter of the present volume.) The Minimal Competencies Panel has worked in liaison with both groups and sees no conflict among the various recommendations.

Nevertheless, each of these two main matters will be discussed further in the remaining sections of this report. These discussions are intended primarily to clarify the panel's recommendations, but partly as a way of passing along some of the good ideas it has collected. The separation of the two matters is certainly not intended to imply that remedial courses should do nothing to convey an appreciation of mathematics, or that techniques are out of place in mathematics appreciation courses.

Mathematics for Coping with Life

The idea that all college graduates should be expected to have acquired a certain familiarity with mathematics rests in part on the well-founded belief that such a familiarity is necessary for effective functioning in contemporary life, and certainly for life in those spheres college graduates are most likely to enter. Indeed, it may be argued convincingly—and has been argued many times—that a modest acquaintance with mathematics is necessary for the successful functioning of almost *any* member of modern society. But any prerequisite for contemporary life in general ought to be, *a fortiori*, something one has a right to expect of all college graduates.

Unfortunately many students manage to enter college without having learned the mathematics needed for coping with everyday life, and a deplorable fraction of them leave college in the same condition. The panel's recommendations—most explicitly Recommendation C—suggest that for such students there should be at least one course where basic mathematical deficiencies may be repaired.

Students entering college with mathematical deficiencies have presumably had opportunities to learn the mathematics, and for them those opportunities did not work. Therefore, *the college remedial course should not be a mere rehash, and certainly not an accelerated one, of the traditional secondary or even elementary course.*

Courses that cover the same old ground in much the same old way tend to be just as uninspiring and unintelligible for these students as the originals, and therefore even less likely to succeed. Students should be able to find even remedial courses fresh, interesting, and significant.

Many courses of this type are being offered, and new ideas are being tested all the time. Several approaches have been described in print (see, for instance, the CUPM booklet *A Course in Basic Mathematics for Colleges,* reprinted in *A Compendium of CUPM Recommendations,* Vol. 1, pp. 256-313), and other reports will surely appear. Here there will be only a sketch to illustrate the type of course that might be considered.

The goals of the course would be to impart mathematical knowledge needed for dealing with most common situations in which deductive reasoning or calculation is needed, and to provide some motivation and preparation for a second course in mathematics that could help the students become educated men and women. It is *not* a goal of the course to teach, once and for all, high school mathematics in its entirety, or to provide background for some standard courses in mathematics or other scientific subjects. (The problem of preparing students for mathematics courses required in their fields is discussed at length in the report of the Committee on Improving Remediation Efforts in the Colleges.)

Students in the course would typically have studied no mathematics for three or four years, and have been bored, mystified, or discouraged by past experiences with mathematics courses. Remedial courses should be taken during a student's *first two* years of college. There should be *no* formal prerequisites.

The course should be relatively brief (twenty to thirty meetings), and should be managed in such a way that students participate actively and receive frequent personal attention. To facilitate this, there should be approximately a fifth as many student assistants as there are students. The first few times the course is offered, the assistants might be mathematics or science majors; later, they should be students who have succeeded in this and at least one further mathematics course.

Equipment might include identical calculators for the students, the assistants, and the instructor. The calculator should have the four basic arithmetical operations, sign changes, squares, square roots, floating decimal, a one-word memory, and very little else. A device for projecting the face of the instructor's calculator on a screen would be useful. There should also be a large collection of advertisements, newspaper and magazine articles, sales and credit agreements, and so on, the interpretation or use of which would require some of the topics listed below. These might be complemented by *reasonable* imaginary examples, but the illustration of no topic should depend entirely on artificial applications. If no genuine examples can be found, why should the topic be included? In some topics, however, a step should be taken beyond the evidently practical.

Students should be supplied with a single page of formulas, sufficient for the whole course.

The grading policy should be compassionate but firm. Tests should be frequent and repeatable at least once. They should be straightforward, but only high scores should be considered passing. Mastery should be recognized irrespective of the number of attempts needed to show it, within limits, but outstanding performance should be recognized. If possible, permanent records of students who need to repeat the course should not show the unsuccessful tries.

One list of topics for such a course is given below. Additions and modifications should be made in response to real-world needs and to experience in offering the course.

1. Positive decimals; conversion of fractions to decimals with the calculator.
2. Pencil-and-paper arithmetic with signed whole numbers.
3. Pencil-and-paper arithmetic with signed fractions. (There should be no three-or-more digit numerators or denominators, except powers of ten.)
4. Calculator arithmetic with signed decimals.
5. Rounding off.
6. Estimation; orders of magnitude.
7. Scientific notation.
8. Units of measurement; elements of the metric system.
9. Percent.
10. What is a formula? What is a function?
11. Times, distance, and rates.
12. Area and volume.
13. What is an algorithm? Flowcharting.
14. Statistics and its dangers.
15. When is an argument correct?
16. Compound interest.
17. Exponential change.

This list should not give rise to hideous visions of workbooks filled with drill exercises. Games, problems of obvious everyday interest, opportunities for creativity, and occasional attention to general problem-solving strategies should contribute to a cheerful and progressive atmosphere and a positive experience.

Mathematics Appreciation

While the panel does not insist that a knowledge of the cultural side of mathematics should be required of all college students, its Recommendations D and E above suggest that attractive and accessible courses dealing especially with that aspect should be offered. This section of the report contains some reasons for this position and some comments on how it might be realized.

Mathematics has played a central role in the development of modern civilization. It has been essential not only to the growth of science and technology, but has had profound effects on philosophy and other forms of thought as well.

There was certainly no doubt in past centuries that every college graduate, to be an educated person, had to know some mathematics. In medieval times, for example, four of the seven traditional liberal arts were largely or wholly mathematical. The importance attached to mathematics was evident in courses of study in the nineteenth century, and this carried over into the twentieth. Now, however, it is possible to graduate from many colleges without any contact with mathematics beyond the most elementary high-school courses.

While high-school mathematics is important, it does tend to emphasize development of skills. The same, unfortunately, may be said of most college courses whose mission is primarily remedial or preprofessional. But an educated, well-informed person should know something about mathematics beyond skills.

To many, the distinction between mathematicians and accountants is not clear. People who are alert and informed about many things, even colleagues in a university, sometimes assume that mathematicians are constantly doing arithmetic and are surprised to hear that there is such a thing as mathematical research. Their experiences with school mathematics left them with the impression that mathematics is ancient and immutable, and consists of rules and formulas for unfortunate school children to memorize.

The great mathematicians do not occupy their rightful place in the public consciousness. In his *New Yorker* article on mathematics (February 19, 1972), Alfred Adler rightly observed that

> . . . it would be astonishing if the reader could identify more than two of the following names: Gauss, Cauchy, Euler, Hilbert, Riemann. It would be equally astonishing if he should be unfamiliar with the names of Mann, Stravinsky, de Kooning, Pasteur, John Dewey. The point is not that the first five are the mathematical equivalents of the second five. They are

not. They are the mathematical equivalents of Tolstoy, Beethoven, Rembrandt, Darwin, Freud. The geometry of relativity—the work of Riemann—has had consequences as profound as psychoanalysis has . . . .

Many college graduates know a great deal of mathematics; most of them have had to take mathematics in preparation for their work. But how many of these, or how many mathematics majors, for that matter, could tell much about Abel or Jacobi? More important, how many of them could comment plausibly on the relation of mathematics to other disciplines?

The point here is not that mathematics and mathematicians should be glorified but that a reasonable perspective on the place of mathematics in the human enterprise should be more widely shared.

A course designed specifically to improve this perspective would ideally give some idea of what sorts of problems mathematicians consider and how such problems are attacked. The object would be to promote mathematical literacy, interpreted to include an awareness among future colleagues in colleges and universities, in business, in industry, in government, and in many other callings of what mathematics is, why it is important, and how it might serve them. Some history should be covered along the way, but a straight course in the history of mathematics is not recommended for this purpose; it can have meaning only if the students already have some understanding of the mathematical ideas whose development is traced.

The course could include, for example, a discussion of the Euler formula for polyhedra—and the names of Euler, Descartes, and Cauchy already would have entered the discussion. An account of non-Euclidean geometry would be appropriate, and provide an occasion for introducing Gauss and Riemann as well as Bolyai and Lobachevski, and for commenting on the element of arbitrariness in mathematical modeling of reality. Neither of these topics requires any high level of algebraic skill. A discussion of the insolubility of the quintic equation might involve more algebra but would refer to the work of Lagrange, Galois, and Abel—and the important idea of mathematical impossibility would have arisen. There are many other topics that bring up important mathematical ideas and events but do not require much background.

Axiomatics, though obviously important, should not be overemphasized. Axiomatic systems should not be presented in detail unless one obtains by their use some interesting results that were not intuitively obvious from the start. Elementary graph theory offers some nice opportunities here, as well as a great variety of easily understood applications. Laborious efforts to prove the obvious can convince people that the whole endeavor is silly.

Applications are appealing to many students and should be included. There are convenient sources of authentic applications of mathematics at every level of difficulty. Applications, however, should not be allowed to upstage the real star of the show, mathematical thought itself. Calculators and computing might have their place in the course, and some time could profitably be spent on the place of computers in modern society. Serious study of computer science, however, is probably best left to other courses.

The course should give students copious evidence that mathematics has not only played a great part in human history, but continues to thrive in the service of other fields and as an independent source of intellectual excitement and aesthetic appeal. Mathematical "current events," such as the solution of the four color problem and the discovery of new large primes should be mentioned. Something might be said about Hilbert's problems and the Fields medals. Carefully selected readings from *Scientific American, The Mathematical Intelligencer,* and similar publications can help.

The choice of faculty for an appreciation course is critical. It is an extraordinary teaching assistant who would have the experience and breadth of outlook to teach such a course. It should usually be taught by senior faculty, and if appropriate faculty cannot be found, the course should not be taught at all. And it is better that it be taught by the right faculty in larger sections than by reluctant or inept instructors in small ones.

The course mentioned in Recommendation E offers further opportunities. It is still too easy for mathematics and science majors to complete their programs without knowing that research is done in mathematics, that mathematics has deep and productive relationships with many fields, and that mathematics has a rich and fascinating history. A mathematics appreciation course for students with good technical proficiency in mathematics can do much to take care of this and be a memorable experience for all concerned.

As has already been said in Recommendation D, *these observations about separate mathematics appreciation courses should apply, to some extent, to all mathematics instruction,* even remedial. In a perfect world every mathematics course would be a mathematics appreciation course. The world, however, is not perfect.

Appendix

The panel began by consulting the pertinent literature; officers of organizations represented in the Council of Scientific Society Presidents or the Conference Board of Mathematical Sciences, and a sample of mathematicians drawn at random from the 1978-1979 *Combined Membership List.* Summaries of the results may be obtained from the chairman of the panel.

A general announcement and appeal for information and ideas also appeared in *Notices of the American Mathematical Society, Change Magazine, The Mathematics Teacher, The Chronicle of Higher Education, The Two-Year College Mathematics Journal, SIAM News,* and *The American Mathematical Monthly.*

From the first two surveys mentioned, the panel learned not much more than that no national organization in this country, the MAA itself not excepted, has ever taken a position on what college graduates *in general* should know of mathematics.

The survey based on the *Combined Membership List* (CML) and the appeal in periodicals, though more productive, did not provide as much unambiguous guidance as the panel had hoped to get. The CML survey yielded 335 usable responses from a thousand questionnaires. 226 were from persons at colleges and universities. Of these, 105 (39.5%) were from institutions where a mathematics requirement for graduation was in force. These 105 respondents were asked about the nature of the requirement, whether they favored it, and whether they thought it was effective. In the great majority of cases (91 or 86.7%) the requirement could be satisfied by one or more courses. Seven of these respondents reported that the requirement could be satisfied by examination; five others said both courses and an examination were required.

One hundred (95.2%) of the 105 said they favored the requirement, and 75 (71.4%) said they thought it was at least partially effective.

The median course requirement, where one existed, was between 3 and 4 semester hours. A specific course or sequence of courses was seldom required; indeed, acceptable courses were remarkably diverse.

The 161 respondents in colleges and universities which had no general mathematics requirement were asked whether they favored such a requirement. In reply, 148 expressed a preference, and of these 104 (70.3%) favored some kind of a requirement.

When the two groups are combined, one finds that 204 of 253 (80.6%) of those college- or university-affiliated mathematicians in the sample who expressed any preference favored some general graduation requirement in mathematics. The panel did not expect this fraction to be so high. (Unfortunately, the questionnaire did not ask for reasons for the preferences expressed.)

All respondents, academic or not, were asked to mark in a forty-item list of mathematical topics those they thought should be required of all college graduates. The following topics were marked by at least half of the respondents:

- Basic arithmetic skills (94.6%)
- Area and volume of common figures (76.4%)
- Linear equations (71.3%)
- Algebraic manipulations (63%)
- Elementary statistics (55.5%)
- Graphing of elementary functions (54.9%)
- Integer and fractional exponents (54.3%)
- Elementary plane geometry (51.9%)

Next in order were: elementary probability (49%), general problem-solving skills (heuristic) (49%), quadratic equations (47.5%), mathematics in business (46.9%), and radicals (43.9%). Computer programming was marked by 33.1%, just after elementary logic (35.5%) and systems of equations (35.2%).

The question about what standard courses should be required elicited a wide variety of answers, many of which were in fact far from standard. College algebra (mentioned by 51 respondents) led the list, and was followed by probability and statistics (47), calculus (45), elementary or intermediate algebra (44), and computer programming or appreciation (30).

About 45% of the respondents accepted an invitation to comment further. Many merely expanded on earlier answers, but some submitted careful statements of their views. These statements, though not easy to summarize, were carefully studied by the panel.

Responses to the appeal in periodicals were interesting too, but they are even less reducible to a brief summary.

The panel met three times and also conducted a voluminous correspondence within itself and with others. It completes this report with high respect for the complexity of the problem, but hopes that its proposals will be of some use in finding solutions.

Panel Members

DONALD W. BUSHAW, CHAIR, Washington State University.

GERALD L. ALEXANDERSON, University of Santa Clara.

ROBERT J. BUMCROT, Hofstra University.

JUANITA J. PETERSON, Laney College.

EDWIN H. SPANIER, University of California, Berkeley.

Mathematics Appreciation Courses

This chapter contains the report of the CUPM Panel on Mathematics Appreciation Courses. The Panel's report was originally published in two parts in the AMERICAN MATHEMATICAL MONTHLY, Vol. 90 (1983): the text appeared on pp. 44-51, while the references appeared in the Center Section. For this reprinting, the report has been re-edited to include the references in the text of the report. Jerome Goldstein, Chair of the Panel, has prepared a brief new preface for this reprinting.

1989 Preface

Mathematicians generally share the view that all well-educated people should be mathematically literate. As a result, "mathematics appreciation courses" continue to be offered with regularity to students in the fine arts, in the humanities, in some social sciences, and in education. All students who receive college-level training in mathematics deserve to have well-conceived courses, centered around significant mathematics. In particular, students should get a glimpse of what it is that attracts mathematicians to their subject.

The CUPM Panel on Mathematics Appreciation Courses emphasized the course objectives rather than the intended audience, and stressed philosophy and teaching strategies rather than specific content. In fact, the comments in the Panel's report, with their heavy emphasis on attitudes and teaching strategies, have universal appeal and can be read with profit by all college mathematics teachers.

The report is as timely now as when it was written.

JEROME A. GOLDSTEIN
Tulane University
March, 1989

Introduction

In 1977 the Committee on the Undergraduate Program in Mathematics (CUPM) established a panel to consider the content of those college and university courses that treat mathematics appreciation for students in the arts and humanities. Such courses are taken by a large number of students, frequently as their last formal contact with mathematics. Yet in most institutions they are given very low priority; they are frequently taught perfunctorily, without a clear set of objectives, by faculty who lack appropriate interest or credentials. Since these courses may play a major role in molding nonscientists' opinions of mathematics and its role in society, CUPM decided that it should call attention to the importance of these courses and offer some suggestions on how they may be organized and taught effectively.

This is the report, approved by CUPM, of the CUPM Panel on Mathematics Appreciation Courses. While the panel has many guidelines and recommendations to offer, it does not feel that a particular selection of topics or teaching strategy should be universally adopted for mathematics appreciation courses. A main goal of such courses is to get students to appreciate the significant role that mathematics plays in society, both past and present. All material presented in such courses should be well motivated and related to the role of mathematics in culture and technology.

Philosophy

The inclusion of a mathematics appreciation course in the undergraduate curriculum is common in the nation's colleges and universities. This trend is a direct result of an underlying belief, held by most mathematicians, that every well-educated person should be mathematically literate. Whether or not a mathematics course is required at a particular institution often depends, among other things, upon the extent to which this belief is shared by the general faculty. But the ultimate success of an "appreciation" course in mathematics should not depend upon mandatory enrollment. Rather, the value and importance of such a course should be directly attributable to the care and understanding with which it is conceived and taught.

If as mathematicians we accept the notion that an educated person should know something about mathematics, then we must also accept the responsibility for conscientiously providing appropriate training. Students in the mathematical, physical, life, and some social sciences, and usually those in business, study mathematics as an inherent part of the undergraduate curriculum. It is not to these students, but rather to majors in the arts, in the humanities, and in certain social sciences that we must direct the mathematics appreciation course. At the outset we must take into account the background and interests of the prospective students. In many cases they have chosen their majors precisely because of a weak or unpleasant mathematical background; a college course that reinforces this negative experience with mathematics certainly cannot be called an appreciation course.

At most institutions the great majority of students in a mathematics appreciation course will have studied less than four years of high school mathematics; moreover, many of these students will have had poor experience in mathematics, or will have had very weak courses. However, high school mathematics study is predominantly concerned with developing skills, and while such skills are of unquestioned importance, they are not necessarily prerequisite to (nor should teaching them be a part of) a mathematics appreciation course.

Among all fundamental academic disciplines, mathematics is perhaps unique in the degree to which it is not understood (or is misunderstood) by students and even faculty from other areas of study. By taking an introductory course in chemistry, history, or psychology, a student is expected to gain an understanding of the general techniques, accomplishments, and goals of the discipline, and will learn to appreciate the work of the contemporary professional practitioner of the subject, sometimes even to the extent of reading the current journals. But an undergraduate major in mathematics is unlikely to have comparable insight into mathematics. Thus the challenge of a mathematics appreciation course is enormous.

The ultimate goal of such courses is defined by our umbrella title—to instill in the student an appreciation of mathematics. For this to occur, students must come to understand the historical and contemporary role of mathematics, and to place the discipline properly in the context of other human intellectual achievement.

From the beginning of recorded history, mathematics has proved to be an indispensable aid to the empirical sciences; the great successes (and failures) of mathematical reasoning in the furtherance of human knowledge are tales begging to be told. Even the direct impact of mathematics on developments in virtually all disciplines is often not realized by the mathematical layman.

But of course, to mathematicians, the subject is more than a tool of applied science, more than a universal language useful for communication and research in other disciplines. Mathematicians see mathematics as an intellectually exciting discipline, one that holds great aesthetic appeal for its practitioners. This idea of mathematics as art is often difficult for nonmathematicians to appreciate, yet is fundamental to understanding the development and role of the subject.

Finally, to appreciate mathematics fully, one must recognize it as a vital, on-going discipline, one that is practiced by a world-wide community of dedicated, sometimes passionate, and frequently brilliant scholars. It is a surprise to many that mathematics is a living, changing, developing subject. A true appreciation of mathematics requires some knowledge of contemporary developments.

The entire mathematical community should be concerned with what view educated, informed people have of mathematics. Thus, courses in mathematics appreciation, while presumably benefitting primarily the students, may also have a long-term positive effect on the discipline itself. Obvious benefits will accrue if leaders in education, industry, business, and government have a better understanding of the nature, role, and importance of contemporary mathematics.

It is a sad commentary on the attitudes of mathematicians that courses in mathematics appreciation frequently command pejorative (albeit informal) labels such as "Math for Poets." Even the supposedly neutral title of "Math for Liberal Arts Students" may convey the connotation of condescension. We must recall that liberal arts education, for a large percentage of the college educated population, is a rigorous, disciplined encounter with the best elements of man's history and culture. The major clientele of the mathematics appreciation courses are liberal arts students, and it is from their ranks that many of society's leaders will emerge.

The panel believes that it is better to describe courses of this type in terms of their objectives rather than their audience. Since the term "mathematics appreciation" brings to mind similar courses in other special fields (e.g., "music appreciation") that generally carry positive connotations with regard to their role in general undergraduate education, and since it conveys concisely what such courses intend to accomplish, standing as a brief reminder of this intention to both teachers and students, the majority of the Panel prefers this title.

Things to Stress

1. *The relationship between mathematics and our cultural heritage.* Students enrolled in mathematics appreciation courses are generally more interested in, as well as more knowledgeable about, the arts and humanities than the sciences; it is natural, therefore, to capitalize on these strengths by appropriate illustrations of the relations between mathematics and music, art, literature, history, and society.

2. *The role of mathematics in history and the role of history in mathematics.* Although the influence of mathematics is often remote, mathematical discoveries have shaped our world in fundamental ways, altering the course of history as well as the way we

live and work. Examples of these influences abound, and should form a major part of any mathematics appreciation course. Historical developments and the evolution of mathematical concepts should be properly emphasized.

3. *The nature of contemporary mathematics.* The mathematics known by most humanities students is ancient mathematics—the geometry of ancient Greece, and the algebra of the early renaissance; not surprisingly, such students have the impression that mathematics is dead. Showing them that it is in fact a vigorous, growing discipline with considerable influence in contemporary society is an important aspect of any course in mathematics appreciation.

4. *The recent emergence of several mathematical sciences.* While the mathematics appreciation course should not be devoted solely to one "modern" area such as statistics, computer science, or operations research, it surely provides an opportunity to use these fields as illustrations of the panoply of contemporary mathematical science.

5. *The necessity of doing mathematics to learn mathematics.* While some parts of the mathematics appreciation course can and should be about mathematics, it is essential that some parts actually engage the students in doing mathematics. Only in this way can they gain a realistic sense of the process and nature of mathematics. Of course it is vitally important that the instructor have appropriate respect for the students' interest and abilities, and that exercises be selected so as to maintain rather than destroy their enthusiasm.

6. *The role of mathematics as a tool for problem solving.* As the language of science and industry, mathematical models are the tool *par excellence* for solving problems. Students in mathematics appreciation courses should be exposed to contemporary mathematical modelling, to gain some appreciation both of its power and its limitations.

7. *The verbalization and reasoning necessary to understand symbolism.* While symbols provide the mathematician and scientist with great power, they obscure the meaning of mathematics from the uninitiated. A great service the teacher of a mathematics appreciation course can provide is to enable students to overcome their fear of symbols, to learn to think through arguments apart from the traditional symbols in which they are expressed.

8. *The existence of a large body of interesting writing about mathematics.* Students in mathematics appreciation courses generally feel comfortable with assignments such as term papers, book reports, and library research because they have become accustomed to these in their humanities courses. There is much good mathematics that can be learned in this way, and assignments can be arranged that utilize these familiar learning tools.

Things to Avoid

1. Do not leave the assignment of an instructor in the mathematics appreciation course until the last minute, and do not assign it on the sole basis of availability. The course requires more planning and preparation than almost any other mathematics course if it is to be successful.

2. Do not simply allow the students to sit back and listen. It is important that they be involved actively. But this need not take the form of daily homework. In fact, drill type assignments should be avoided. The involvement could take the form of projects, papers, book reports, "discovering" mathematics in class, participating in class discussions.

3. Do not over-emphasize the history of mathematics. While the history of mathematics could and should be used to enliven the topics covered, a student who knows (and cares) nothing about a mathematical topic is not likely to be interested in its history.

4. Do not stress remedial topics. While many of the students in a mathematics appreciation course may need remedial work, any such material that is covered must be presented as part of a topic that fits into the scope of the course as a whole.

5. Do not make a fetish of rigor; in particular do not prove things that are self-evident to the students. For example, a rigorous presentation of the real numbers in which one proves the uniqueness of zero is entirely inappropriate in courses of this type.

6. Do not cover topics you do not yourself find interesting and important. It is hard to fool these students, and if the teacher does not care, they will not see why they should.

7. Do not be condescending. While the students in such courses may not be mathematically inclined, this does not mean that they are unintelligent. Many who take mathematics appreciation courses are outstanding, creative students who have simply concentrated in the nonquantitative areas of the curriculum. The attitude of the teacher can help either to open or to close their minds to the material.

8. Do not cover topics which you cannot relate in some way to ideas familiar to the students. Clock arithmetic and symbolic logic, for example, are of little value to mathematics appreciation courses unless

you can find applications the students can appreciate and understand.

9. Do not make the course too easy. The material should not be way over the heads of the students, but it should not be trivial either.

10. Do not accept anyone else's blueprint for a mathematicians appreciation course. If you can communicate, in your own way, why you believe that mathematics is beautiful and important, the course will fulfill its purpose.

Course Organization

There are nearly as many ways to teach a course in mathematics appreciation as there are teachers of these courses. While some strategies will work superbly in some contexts, none can be recommended for all; the teacher's enthusiasm for what is being done as well as the appropriateness of the strategy for the students in the course is generally more important than the actual strategy adopted. Nevertheless, to encourage flexibility, we list below some approaches to teaching mathematics appreciation that have been effective in certain contexts.

1. A sampler approach, featuring a variety of more or less independent topics. The advantage of this method is that it covers many areas without requiring a sustained continuity of interest; students who fall behind or simply fail to comprehend one topic always know that they have a chance for a fresh start in a few days. The disadvantage is that of all survey courses: not enough time spent on any one thing to ensure long-term learning.

2. A single-thread approach, built around a common theme, for example, 2 x 2 matrices, or algorithms, or patterns of symmetry. Doing this takes careful planning, and runs the risk of alienating some of the class who find the thread incomprehensible. But it guarantees a solid example of the intellectual coherence that is so much a part of contemporary mathematics, that ideas arising in one context find applications in others, and that a common abstract structure underlies them all.

3. A Socratic approach, in which the instructor works carefully to let the students develop their own reasoning. This works well in small classes with a highly-motivated instructor. While the content of such courses is hard to guarantee in advance, the achievement for students who are able to think for themselves, perceiving patterns where others simply see chaos, is a worthy objective for a course in mathematics appreciation.

Examples of Topics

The topics available for courses in mathematics appreciation are as diverse as mathematical science itself. Standard textbooks offer a rather traditional assortment of topics: probability, graph theory, finite difference equations, computers, matrices, statistics, exponential growth, set theory, and logic seem to dominate. But there are numerous other themes that can be used for large or small components of courses. Here are a few of the many possible examples.

1. Understanding how to use the buttons on a pocket calculator. It used to be that the number e was a complete mystery to those who had not studied calculus, and that "sin" had for humanities students more the connotation of theology than of mathematics. But no more. Virtually everyone has, or has seen, inexpensive hand calculators with buttons that perform operations involving exponential, trigonometric, and basic statistical functions. Teaching a class what these buttons do is an exciting new way to explore some traditional parts of classical mathematics.

   REFERENCES: See the handbooks for various calculators.

2. Tracing the modern descendants of classical mathematical ideas can illustrate the power of mathematics to influence the real world, as well as its remoteness from it. For example, classical Greek geometry involving conic sections led to models for planetary motion, and ultimately to the possibility of space flight. And probability, which had its origins in seventeenth-century discussions about gambling, now dominates actuarial and fiscal policy, influencing government and corporate budgets, thus affecting the level of interest, of unemployment, and the health of the entire economy.

   REFERENCES: Much of this is in standard textbooks. Morris Kline's *Mathematics in Western Culture* and George Pólya's *Mathematical Models in Science* are helpful sources.

3. Connecting mathematics with Nobel prizes. Nobel prizes are not given in mathematics (and the apocryphal reasons for this are quite amusing). But the work that led to Nobel prizes (e.g., of Libby, of Arrow, of Lederberg, and others) often has an intrinsically-mathematical basis. The study of this scientific work provides an opportunity to show how mathematics is important in the most profound discoveries of modern science.

   REFERENCES: Libby's work is briefly discussed in several elementary texts on ordinary differential equations, e.g., in *Differential Equations with Applications and Historical Notes* by George F. Simmons. For some work of Arrow see Edward Bender, *An Introduction to Mathematical Modeling*, Wiley, New York, 1978. Lederberg published an article in this *Monthly* entitled "Hamilton circuits of convex trivalent polyhedra (up to 18 vertices)" in Vol. 74, pp. 522-527.

4. Applying exponential growth models. The applications of traditional topics from elementary mathematics can often be explored more fully than has usually been the case. Exponential growth and decay models provide a striking example. Simple non-calculus approaches to models of growth provide a basis for discussion not only of interest and inflation, but also of such things as radio-carbon dating, cooling and heating of houses, population dynamics, strategies for controlling epidemics, and even detection of art forgeries.

   REFERENCES: See any modern text on ordinary differential equations. A particularly good one is Martin Braun, *Differential Equations and Their Applications*, Springer-Verlag, New York, 1975.

5. Relating traditional mathematics to new applications. A discussion of beginning probability theory can quickly lead to a treatment of the Hardy-Weinberg law of genetics and a calculation of the probability of winning state lotteries. An introductory treatment of statistics can quickly lead to a discussion of political polls, the design and interpretation of surveys and of related decision-making problems. Modern applications of elementary network theory include recent work in computational complexity and almost unbreakable codes.

   REFERENCES: The Hardy-Weinberg law appears in several texts on finite mathematics, e.g., *Applied Finite Mathematics* by Anton and Kolman. *How to Lie with Statistics* by Darrel Huff and Irving Geis, Norton, 1954, and other texts contain situational mathematics which can be discussed according to the interests of the audience. For the two topics mentioned last, see *Scientific American*, Jan. 1978 and Aug. 1977.

6. Introducing problems involving decision making. There are many situations described by elementary mathematics in which one must choose "rationally" among possible options. One can discuss quantifying risk and uncertainty, fair division schemes, applications of network flows, pursuit and navigation problems, game theory, and numerous other topics. Political science is full of unexpected but usually interesting topics, including Arrow's theorem and its offshoot theories of voting, the recently discovered problems associated with apportionment of legislatures, and strategies of fair voting in multiple candidate elections.

   REFERENCES: See Bender, *An Introduction to Mathematical Modeling*, Wiley, New York, 1978; the articles by William Lucas in Vol. 2 of the forthcoming *Modules in Applied Mathematics* (Springer-Verlag, New York); M. Balinski and H.P. Young, *Proc. Nat. Acad. Sci. U.S.A.* 77 (January 1980) 1-4; H. Hamburger, *J. Math. Sociology* 3 (1973) 27-48; David Gale, UMAP Module 317, 1978; W. Stromquist, this *Monthly* 87 (1980) 640-644; and George Minty's article in M.D. Thompson, ed., *Discrete Mathematics and its Applications* (Indiana University, Bloomington, 1977).

7. Exploring the powers and limitations of mathematical models. Each of the modern social sciences abounds with applications of elementary mathematics. All of the examples mentioned above, and many more, involve the use of mathematical models. Sometimes these models are quite accurate and sometimes they are not. But even in the latter case the model can help clarify one's thinking about the underlying problem. An example of this use of mathematical modelling is the prisoner's dilemma argument of game theory and its possible connection with U.S.-U.S.S.R. relations.

Two-Year Colleges

Many courses that ought to follow the "mathematics appreciation" philosophy are taught in two-year colleges. Innovative approaches and curriculum development by some two-year faculty are reflected by their texts and articles in this area. Although the preceding sections of this report are applicable to mathematics appreciation courses in all colleges, this separate section appears because of the special problems created in two-year colleges by generally heavy teaching loads, by staffing in some cases by faculty whose mathematical experiences are not sufficient to make them comfortable with the broad range of topics demanded by these courses, and by the regrettable frequency of administrative procedures which allow students needing remediation to enroll in these courses. The following suggestions may help to overcome these impediments to two-year college implementation of the goals of mathematics appreciation courses.

1. When there is a choice among faculty members for assignment to the mathematics appreciation course, only those having a broad range of mathematical experiences and expressing interest in the course

should teach it. Mathematics program administrators should provide extra guidance to faculty teaching this course for the first time. In the two-year college there will usually be one text used by all teachers, often supplemented by a reading list and/or other texts; a description of special uses of these materials, as well as sample course outlines, supplementary and classwork materials, and tests, will be helpful. Entrance and/or exit requirements may be matters of policy and should be explained. Lists of applicable resource material owned by the school should be provided to new teachers, along with knowledge of the schools' firm rental policies.

2. Special attention should be paid to the needs of this course by the library, audio-visual, and computer facilities. The mathematics program administrator should be sure these courses are adequately supported.

3. Since mathematics appreciation courses, properly taught, take an enormous amount of preparation time, any load relief possible would be appropriate. In a suitable lecture room, the course can be effectively taught to "double" sections of 60-90 students if doing so would leave the teacher several hours more preparation time each week. (Such a load might be counted as two or three sections, corresponding to the grading load.)

4. Sharing materials and ideas and perhaps team-teaching would be reasonable for mathematics appreciation courses. One teacher at a school might be most qualified for teaching, say, a computer unit, and might "rotate" across several sections. Many more "hand-out" materials seem to be necessary for mathematics appreciation courses than for traditional courses; these might be used by several teachers in a given term, or re-used in succeeding terms. Faculty teaching mathematics appreciation courses seem to enjoy sharing materials and methods.

5. In-course remediation should be avoided. If students are enrolled who cannot handle elementary operations at the level needed for the work of the course, a "math lab" facility might be used to design and administer individual remediation programs. It cannot be over-emphasized that a mathematics appreciation course cannot fulfill its goal if it degenerates into the teaching of arithmetic computations or pre-algebra skills, or if it is limited to a topic such as "consumer mathematics."

6. A large proportion of students enter two-year colleges with little realistic expectation concerning majors. Many of these students have had poor experiences with mathematics and, if there is a general education mathematics requirement which may be satisfied by either a mathematics appreciation course or a pre-calculus/calculus course, they will often elect the mathematics appreciation course. Well into a successful term, the student may begin to think realistically about mathematics requirements of various university majors. Since most majors outside the humanities will necessitate at least some mathematics at a technical level rarely achieved in the typical two-year college mathematics appreciation course, an important service of this course can be to channel these students back into regular sequence mathematics courses.

Without violating the spirit of a mathematics appreciation course, it is possible to include a topically organized unit requiring the review and use of elementary algebra and graphing techniques; this may give the student a successful experience in doing mathematics that serve as encouragement to return to regular sequence mathematics courses. (A linear programming unit, for example, requires the students to review or acquire facility with graphing and algebra techniques. Many of the topics suitable for a mathematics appreciation course can be handled in this way.) Students with the experience will frequently place higher in the sequence courses than they would have upon original enrollment, and will go on as solid, though late-blooming, students.

Acknowledgements

Many people, including nonmathematicians, responded to the Panel's request for information and advice. An open meeting of the panel at the MAA meeting in Biloxi, Mississippi, in January 1979, was attended by more than sixty people who had much to contribute. The panel particularly wishes to thank Professors Henry Alder, Dorothy Bernstein, Donald Bushaw, William Lucas, David Penney, William M. Priestley, David Roselle, and Alice Schafer for their helpful comments.

Panel Members

JEROME A. GOLDSTEIN, CHAIR, Tulane University.
BETTYE ANNE CASE, Florida State University.
JOHN CONWAY, Indiana University.
RICHARD J. DUFFIN, Carnegie-Mellon University.
ELAINE KOPPELMAN, Goucher College.
KENNETH REBMAN, California State University at Hayward.
LYNN ARTHUR STEEN, St. Olaf College.
JAMES VINEYARD, Laney College.

Reference Section

Films

Since students in mathematics appreciation courses frequently have little experience in sustaining interest in regular mathematics lectures, it is usually appropriate in these courses to provide a variety of class activities. Films are a useful but under-utilized medium for mathematics instruction generally. They are especially useful for the mathematics appreciation course.

The following selection of films about mathematical subjects features those that are suitable for lay audiences. (Distributor addresses are listed at the end of the list.) Further information on these and other films is available in the booklet *Annotated Bibliography of Films and Videotapes for College Mathematics* by David Schneider (M.A.A., 1980).

A Non-Euclidean Universe. (1978; 25 Min; Color). University Media.

A Time for Change—The Calculus. (1975; 25 Min; Color) University Media.

Accidental Nuclear War. (1976; 8 Min; Color) Pictura Film.

Adventures in Perception. (1973; 22 Min; Color) BFA Educational Media. (Reviews: Amer. Math. Monthly 84 (1977) 582.)

An Historical Introduction to Algebra. Modern Film Rentals.

An Introduction to Feedback. (1960; 11 Min; Color) Encyclopedia Britannica Educational Corporation.

Anatomy of Analogy. (25 Min; BW) Open University.

Area and Pi. (1969; 10 Min; Color) Modern Film Rentals.

Auto Insurance. (1976; 8 Min; Color) Pictura Film.

Caroms. (1971; 9 1/2 Min; Color) International Film Bureau. (Reviews: Amer. Math. Monthly 82 (1975) 417; Math Teacher 66 (1973) 51.)

Central Perspectivities. (1971; 13 1/2 Min; Color) International Film Bureau. (Reviews: Amer. Math. Monthly 82 (1975) 419; Math. Teacher (1972) 733.)

Central Similarities. (1966; 10 Min; Color) International Film Bureau. (Reviews: Amer. Math. Monthly 82 (1975) 418; Math. Teacher (1972) 643-644.)

Challenge in the Classroom. (1966; 55 Min; Color) Modern Film Rentals.

Circle Circus. (1979; 7 Min; Color) International Film Bureau.

Common Generation of Conics. (4 Min; Color). Educational Solutions.

Complex Numbers. (1978; 25 Min; Color). University Media.

Computer Perspective. (1972; 8 Min; Color) Pyramid Films.

Congruent Triangles. (1978; 7 Min; Color). International Film Bureau.

Conic Sections. (1968; 11 Min; Color) BFA Educational Media.

Conics. (1979; 10 Min; Color) Wards Modern Learning Aids.

Conics. (1978; 25 Min; Color). University Media.

Constructing an Algorithm. (25 Min; BW) Open University.

Cosmic Zoom. McGraw-Hill Films.

Curves. (1968; 17 Min; Color) A.I.M.S. (Reviews: Amer. Math. Monthly 83 (1976) 71-72; Math. Teacher 64 (1971) 525.)

Curves of Constant Width. (1971; 16 Min; Color) International Film Bureau. (Reviews: Amer. Math. Monthly 78 (1971) 539; Math. Teacher 65 (1972) 234.)

Cycloidal Curves or Tales From the Wanklenberg Woods. (1974; 22 Min; Color) Modern Film Rentals.

Dance Squared. (1963; 4 Min; Color) International Film Bureau. Review: Math. Teacher 64 (1971) 627.)

Dihedral Kaleidoscopes. (1966; 13 Min; Color) International Film Bureau. (Review: Math. Teacher 66 (1973) 51.)

Dimension. (1970; 13 Min; Color) A.I.M.S. (Reviews: Amer. Math. Monthly 83 (1976) 71-72; Math. Teacher 64 (1971) 525.)

Donald in Mathmagicland. (1960; 26 Min; Color) Walt Disney Educational Media Company.

Dr. Posin's Giants: Isaac Newton. Indiana University Audiovisual Library.

Dragon Fold...And Other Ways to Fill Space. (1979; 7 1/2 Min; Color) International Film Bureau.

Equidecomposable Polygons. (10 1/2 Min; Color) International Film Bureau. (Reviews: Amer. Math. Monthly 82 (1975) 687-688; Math. Teacher 65 (1972) 734.)

Errors That Die. (25 Min; BW) Open University.

Flatland. (1965; 12 Min; Color) McGraw-Hill Films. (Review: Math. Teacher 64 (1971) 44-45.)

Functions and Graphs. (1978; 25 Min; Color) University Media.

Geodesic Domes: Math Raises the Roof. (1979; 20 Min; Color) David Nulsen Enterprises.

Geometric Vectors—Addition. (1971; 17 Min; Color) International Film Bureau. (Review: Amer. Math. Monthly 82 (1975) 420.)

Geometry: Inductive and Deductive Reasoning. (1962; 12 1/2 Min; Color) Coronet Films.

Good for What? (25 Min; BW) Open University.

Gottingen and New York. (1966; 43 Min; Color) Modern Film Rentals.

How Far is Around? (1979; 7 1/2 Min; Color) International Film Bureau.

Inferential Statistics, Part I: Sampling and Estimation. (1977; 19 Min; Color) Media Guild.

Inferential Statistics, Part II: Hypothesis Testing. (1977; 25 Min; Color). Media Guild.

Infinity. (1972; 17 Min; Color) A.I.M.S. (Review: Amer. Math. Monthly 83 (1976) 71-72.)

Inversion. (12 Min; Color) International Film Bureau. (Reviews: Amer. Math. Monthly 83 (1976) 71; Math. Teacher (1972) 644.)

Isaac Newton. (1959; 13 1/2 Min; Color) Coronet Films.

Isn't That the Limit! (1980; 17 Min; Color) David Nulsen Enterprises.

Isometries. (1967; 26 Min; Color) International Film Bureau. (Review: Math. Teacher 66 (1973) 51-52.)

Iteration and Convergence. (1978; 25 Min; Color) University Media.

John von Neumann, A Documentary. (1966; 63 Min; BW) Modern Film Rentals. (Review: Amer. Math. Monthly 75 (1968) 435.)

Journey to the Center of a Triangle. (1977; 8 1/2 Min; Color) International Film Bureau.

Let Us Teach Guessing. (1966; 61 Min; Color) Modern Film Rentals. (Review: Amer. Math. Monthly 75 (1968) 219.)

Limit Curves and Curves of Infinite Length. (1979; 14 Min; Silent; Color) International Film Bureau.

Limit Surfaces and Space Filling Curves. (1979; 10 1/2 Min; Silent; Color) International Film Bureau.

Limits. (25 Min; BW) Open University.

Linear Programming. (1969; 9 Min; Color) Macmillan Films.

Look Again. (1970; 15 Min; Color) A.I.M.S. (Reviews: Amer. Math. Monthly 83 (1976) 71-72; Math. Teacher 64 (1971) 525.)

Love Song. (1976; 11 Min; Color) Pictura Film.

Mathematical Curves. (1977; 10 Min; Color) Churchill Films.

Mathematical Induction. (1960; 62 Min; Color) Modern Film Rentals.

Mathematical Induction. (1978; 25 Min; Color) University Media.

Mathematical Peep Show. (1961; 11 Min; Color) Encyclopedia Britannica Educational Corporation. (Review: Math. Teacher 64 (1971) 625.)

Mathematician and the River. (1959; 19 Min; Color) No distributor.

Mathematics of the Honeycomb. (1964; 13 Min; Color) Moody Institute of Science. (Review: Math. Teacher 64 (1971) 334.)

Matrices. (9 Min; Color) Macmillan Films.

Matrioska. Indiana University Audiovisual Library.

Maurits Escher, Painter of Fantasies. (1970; 26 1/2 Min; Color) Coronet Films. (Review: Amer. Math. Monthly 83 (1976) 495.)

Mean, Median, Mode. McGraw-Hill Films.

Modelling Drug Therapy. (1978; 25 Min; Color). University Media.

Modelling Pollution. (1978; 25 Min; Color). University Media.

Modelling Surveys. (1978; 25 Min; Color). University Media.

Modmath. (14 1/2 Min; Color) International Film Bureau.

Mr. Simplex Saves the Aspidistra. (1966; 33 Min; Color) Modern Film Rentals.

Networks and Matrices. (1978; 25 Min; Color) University Media.

New Worlds From Old. (1975; 25 Min; Color) University Media.

Newton's Equal Areas. (1968; 8 Min; Color) International Film Bureau. (Reviews: Amer. Math. Monthly 79 (1972) 1054; Math. Teacher 63 (1970) 449.)

Nim and Other Oriented Graph Games. (1966; 63 Min; BW) Modern Film Rentals.

Notes on a Triangle. International Film Bureau. (Review: Math. Teacher 63 (1970) 363.)

Numbers Now and Then. (1975; 25 Min; Color) University Media.

Orthogonal Projection. (1965; 13 Min; Color) International Film Bureau. (Reviews: Amer. Math. Monthly 82 (1975) 419-420; Math. Teacher (1972) 643.)

Paradox Box. Scientific American.

Pits, Peaks, and Passes (Part 1). (1966; 48 Min; Color) Modern Film Rentals.

Plateau's Problem. A Film by Sr. Rita Ehrmann.

Points of View: Perspective and Projection. (1975; 25 Min; Color) University Media.

Possibly So, Pythagoras. (1973; 14 Min; Color). International Film Bureau. (Review: Math. Teacher 64 (1971) 626.)

Powers of Ten. (1978; 9 Min; Color) Pyramid Films. (Review: Math. Teacher (1979) 388.)

Predicting at Random. (1966; 43 Min; Color) Modern Film Rentals.

Probability. (12 Min; Color) McGraw-Hill Films.

Professor George Pólya and Students, Parts I and II. (1972; 60 Min; Color) University Media.

Professor George Pólya Talks to Professor Maxim Bruckheimer. (1972; 60 Min; Color) University Media.

Projective Generation of Conics. (16 Min; Color) International Film Bureau. (Reviews: Amer. Math. Monthly 82 (1975) 538-539; Math. Teacher 66 (1973) 51.)

Quaternions: A Herald of Modern Algebra. (1975; 25 Min; Color) University Media.

Rational Numbers and the Square Root of 2. (1978; 25 Min; Color) University Media.

Regular Homotopies in the Plane: Part I: (1975; 14 Min; Color); *Part II:* (1975; 18 1/2 Min; Color) International Film Bureau. (Review: Amer. Math. Monthly 85 (1978) 212.

Root Two: Geometry or Arithmetic? (1975; 25 Min; Color) University Media.

Sampling. (25 Min; BW) Open University.

Sets, Crows, and Infinity. (12 Min; Color) BFA Educational Media.

Shaking the Foundations. (1975; 25 Min; Color) University Media.

Shapes of the Future: Some Unsolved Problems in Geometry: Part I: Two Dimensions (1975; 22 Min; Color); *Part II: Three Dimensions* (1970; 21 Min; Color) Modern Film Rentals. (Review: Amer. Math. Monthly 79 (1972) 1052-1053.)

Sierpinski's Curve Fills Space. (1979; 4 1/2 Min; Color) International Film Bureau.

Similar Triangles. (1976; 7 1/2 Min; Color) International Film Bureau.

Space Filling Curves. (1975; 25 1/2 Min; Color) International Film Bureau. (Review: Math. Teacher 69 (1976) 164-165.)

Sphere Eversions. (1979; 7 1/2 Min; Silent; Color) International Film Bureau.

Spheres. International Film Bureau.

Statistics At A Glance. (1972; 28 Min; Color) Media Guild. (Review: Amer. Math. Monthly 82 (1975) 312.)

Statistics and Probability I. (15 Min; BW) Open University.

Statistics and Probability II. (25 Min; BW) Open University.

Statistics and Probability III. (25 Min; BW) Open University.

Symbols, Equations and the Computer. (1978; 25 Min; Color) University Media.

Symmetries of the Cube. (1971; 13 1/2 Min; Color) International Film Bureau. (Review: Math. Teacher 65 (1972) 733.)

The Algebra of the Unknown. (1975; 25 Min; Color) University Media.

The Binomial Theorem. (1978; 25 Min; BW) University Media.

The Butterfly Catastrophe. (1979; 4 1/2 Min; Silent; Color) International Film Bureau.

The Delian Problem. (1975; 25 Min; Color) University Media.

The Dot and the Line. Indiana University Audiovisual Library.

The Geometry Euclid Didn't Know. (1979; 16 Min; Color) David Nulsen Enterprises. (Reviews: Amer. Math. Monthly 86 (1979) 600; Math. Teacher (1979) 300.)

The Great Art—Solving Equations. (1975; 25 Min; Color) University Media.

The Hypercube: Projections and Slicing. (1978; 12 Min; Color) Banchoff-Strauss Productions.

The Kakeya Problem. (1962; 60 Min; Color) Modern Film Rentals.

The Majestic Clockwork. (1974; 52 Min; Color) Time Life Multimedia.

The Marriage Theorem, Parts I & II. (1974; 46 Min. and 47 Min; BW) Modern Film Rentals.

The Music of the Spheres. (1974; 52 Min; Color) Time Life Multimedia.

The Nature of Digital Computing. (25 Min; BW) University Media.

The Perfection of Matter. McGraw-Hill Films.

The Search for Solid Ground. (1963; 62 Min; BW) Modern Film Rentals.

The Seven Bridges of Königsberg. (1965; 4 Min; Color) International Film Bureau.

The Structure of a Computer. (25 Min; BW) Open University.

Topology. (1972; 9 Min; Color) Macmillan Films.

Topology. (1966; 30 Min; BW) Modern Film Rentals. (Review: Amer. Math. Monthly 75 (1968) 790.)

Topology: Some Historical Concepts. (21 3/4 Min; Color) Richard Cline Film Productions.

Trio for Three Angles. (8 Min; Color) International Film Bureau.

Turning a Sphere Inside Out. (1976; 23 Min; Color) International Film Bureau. (Reviews: Amer. Math. Monthly 86 (1979) 511-512; Math. Teacher 70 (1977) 55.)

View From the People Wall. (1964; 14 Min; Color) Encyclopedia Britannica Educational Corporation.

Weather by the Numbers. University of Indiana Audiovisual Library.

What is a Limit? (25 Min; BW) Open University.

What is a Set? Part I & II. (1967; 15 Min; Color) Modern Film Rentals. (Review: Amer. Math. Monthly 75 (1968) 324.)

What is Mathematics and How Do We Teach it? (1966; 60 Min; BW) Modern Film Rentals.

Zooms on Self-Similar Figures. (1979; 8 Min; Color) International Film Bureau.

ADDRESSES OF DISTRIBUTORS:

A.I.M.S., 626 Justin Avenue, Glendale, CA 91201.

International Film Bureau, 332 South Michigan Avenue, Chicago, IL 60604.

McGraw-Hill Films, McGraw-Hill Book Company, 330 West 42nd Street, New York, NY 10036.

Modern Film Rentals, 2323 New Hyde Park Road, New Hyde Park, NY 11040.

Moody Institute of Science, 12000 East Washington Boulevard, Whittier, CA 90606.

Indiana University Audiovisual Library, Bloomington, IN 47401.

Ward's Modern Learning Aids Division, P.O. Box 1712, Rochester, NY 14603.

University Media, 118 South Acacia, Box 881, Solana Beach, CA 92075.

Pyramid Films, P.O. Box 1048, Santa Monica, CA 90406.

Educational Solutions, Inc., 80 Fifth Avenue, New York, NY 10011.

Banchoff-Strauss Productions, Inc., P.O. Box 2430, East Side Station, Providence, RI 02906.

Macmillan Films, Inc., 34 MacQuesten Parkway South, Mt. Vernon, NY 10550.

Encyclopedia Britannica Educational Corporation, 425 North Michigan Avenue, Chicago, IL 60611.

David Nulsen Enterprises, 3211 Pico Boulevard, Santa Monica, CA 90405.

The Media Guild, 118 South Acacia, Box 881, Solana Beach, CA 92075.

Coronet Films, 65 East South Water Street, Chicago, IL 60601.

Pictura Film Co., 111 8th Avenue, New York, NY 10011.

BFA Educational Media, 2211 Michigan Avenue, P.O. Box 1795, Santa Monica, CA 90406.

Time-Life Multimedia, Time and Life Building, New York, NY 10020.

Churchill Films, 662 North Robertson Boulevard, Los Angeles, CA 90069.

Walt Disney Educational Media Company, 500 South Buena Vista Street, Burbank, CA 91521.

Contemporary-McGraw Hill Films, 1221 Avenue of the Americas, New York, NY 10020.

Classroom Aids

Certain topics treated in mathematics appreciation courses are particularly amenable to demonstration with physical or geometric devices. Useful exhibits can often be seen at NCTM meetings. A list of major suppliers of mathematics classroom devices is given below:

The Math Group, Inc., 396 East 79th Street, Minneapolis, MN 55420.

Unique puzzles and card games. Designed for elementary school enrichment, many are flexible enough to be of interest to adults as well.

Lano Company, 9001 Gross Road, Dexter, MI 48130. (313-426-4860).

Mathematical visual aids (solids, transparencies, graphing aids) together with various games and probability models.

Math Shop, Inc., 5 Bridge Street, Watertown, MA 02172.

A full range of curriculum enrichment material for elementary, junior and senior high school mathematics. Includes several games and puzzles of use in introductory college mathematics.

International Film Bureau, Inc., 332 South Michigan Avenue, Chicago, IL 60604.

The major American distributor of mathematics films. The current list includes nearly 50 films at the high school and college level, each of which may be either purchased or rented.

Yoder Instruments, East Palestine, OH 44413. (216-426-3612; 216-426-9580).

Two unique geometric construction sets for plane and solid geometry.

Cuisenaire Co. of America, Inc., 12 Church Street, New Rochelle, NY 10805.

Sensory apparatus designed for elementary schools, some of which (e.g., geoboards, polyhedral structures) would be suitable to courses in mathematics appreciation.

LaPine Scientific Company, Department B43, 6005 South Knox Avenue, Chicago, IL 60629; 373 Chestnut Street, Norwood, NJ 07648; 920 Parker Street, Berkeley, CA 94710; Box 95, Postal Station U., Toronto, Canada M8Z 5M4.

An extensive offering of models, teaching aids, games, and audio-visual materials for elementary, high school and beginning college mathematics. Mathematics catalogue exceeds 100 pages.

Geyer Instructional Aids Co., Inc., P.O. Box 7306, Fort Wayne, IN 46807.

A large collection of models, games, books and classroom aids for high school and elementary college courses.

Creative Publications, 3977 East Bayshore Road, P.O. Box 10328, Palo Alto, CA 94303. (415-968-3977).

An attractive 100-page catalogue of mathematical books, models, games, posters, construction sets, and puzzles for all grade levels. The premier source for mathematics enrichment material.

Inquiry Audio-Visuals, 1754 West Farragut Avenue, Chicago, IL 60640.

Filmstrips and transparencies for high school algebra topics.

Educational Audio Visual, Inc., Pleasantville, NY 10570.

Transparencies and games for algebra, geometry, calculus and statistics.

W.H. Freeman and Company, 660 Market Street, San Francisco, CA 94104.

Offprints of Scientific American articles.

David Nulsen Enterprises, 3211 Pico Boulevard, Santa Monica, CA 90405.

Three 15-minute, 16mm color films ("Curves," "Dimension," "Look Again") for rent or purchase.

Popular Science Audio-Visuals, Inc., 5235 Ravenswood Avenue, Chicago, IL 60640.

Filmstrips and overhead transparencies, principally in high school mathematics (geometry, algebra, elementary functions).

Time Life Films, 43 West 16th Street, New York, NY 10011. (212-691-2930).

Numerous BBC-produced 20-minute 16mm B & W films on topics ranging from inequalities to matrices.

Edmund Scientific Company, 1985 Edscorp Building, Barrington, NJ 08007. (609-547-3488).

References

Since many of the topics that arise in mathematics appreciation courses occur nowhere else in the mathematics curriculum, it is quite important that instructors be aware of the expository literature of mathematics that treats its relations to science and society. Student term papers in courses on mathematics appreciation typically tax the instructor's knowledge of the literature more than any other course in the mathematics curriculum.

To aid instructors of mathematics appreciation courses, we prepared a list of major references that would be suitable for background reading, and as sources for special projects. This list does not include textbooks, partly because we do not wish to endorse some books over others, and partly because texts go in and out of print much more rapidly than the reference classics.

Survey Monographs

Boehm, George A.W. *The New World of Mathematics.* The Dial Press, New York, 1959.

Courant, Richard and Robbins, Herbert. *What is Mathematics?* Oxford University Press, New York, 1941.

Dantzig, Tobias. *Number, the Language of Science,* 4th ed. Free Press, New York, 1967.

Goarding, Lars. *Encounter with Mathematics.* Springer-Verlag, New York, 1977.

Herstein, I.N. and Kaplansky, I. *Matters Mathematical.* Harper and Row, New York, 1974.

Khurgin, Ya. *Did You Say Mathematics?* MIR Publishers, Moscow, Russia, 1974.

Kline, Morris. *Mathematics: The Loss of Certainty.* Oxford University Press, New York, 1980.

Pedoe, Daniel. *The Gentle Art of Mathematics.* Macmillan, Riverside, New Jersey, 1958, 1963.

Rademacher, Hans and Toeplitz, Otto. *The Enjoyment of Mathematics.* Princeton University Press, Princeton, New Jersey, 1957.

Sawyer, W.W. *Introducing Mathematics,* 4 vols. Penguin Books, New York, 1964-70.

Singh, Jagjit. *Great Ideas of Modern Mathematics, Their Nature and Use.* Dover, New York, 1959; Hutchinson and Company, London, England, 1972.

Stein, Sherman K. *Mathematics, The Man-made Universe: An Introduction to the Spirit of Mathematics,* Third Edition. W.H. Freeman, San Francisco, California, 1976.

Steinhaus, Hugo. *Mathematical Snapshots,* 2nd ed. Oxford University Press, New York, 1969.

Stewart, Ian. *Concepts of Modern Mathematics.* Penguin Books, New York, 1975.

Whitehead, Alfred North. *An Introduction to Mathematics.* Oxford University Press, New York, 1958.

Collections of Essays

Kline, Morris, ed. *Mathematics in the Modern World.* W.H. Freeman, San Francisco, California, 1968.

Kline, Morris, ed. *Mathematics: An Introduction to Its Spirit and Use.* W.H. Freeman, San Francisco, California, 1979.

LeLionnais, F., ed. *Great Currents of Mathematical Thought,* 2 vols. Dover, New York, 1971.

Messick, David M., ed. *Mathematical Thinking in Behavioral Sciences.* W.H. Freeman, San Francisco, California, 1968.

National Research Committee on Support of Research in the Mathematical Sciences (COSRIMS). *The Mathematical Sciences– A Collection of Essays.* MIT Press, Cambridge, Massachusetts, 1969.

Newman, James R., ed. *The World of Mathematics,* 4 vols. Simon and Schuster, New York, 1956-60.

Saaty, Thomas L. and Weyl, F. Joachim, eds. *The Spirit and Uses of the Mathematical Sciences.* McGraw-Hill, New York, 1969.

Schaaf, William L., ed. *Our Mathematical Heritage,* New, Revised Edition. Collier Books, New York, 1963.

Steen, Lynn Arthur, ed. *Mathematics Today: Twelve Informal Essays.* Springer-Verlag, New York, 1978.

Nature of Mathematics

Adler, Alfred. "Reflections–mathematics and creativity." *New Yorker* 47 (February 19, 1972) 39-45.

Bronowski, Jacob. "The music of the spheres." in J. Bronowski, *The Ascent of Man.* Little, Brown, and Company, Boston, Massachusetts, 1973, pp. 154-187.

Bronowski, Jacob. "The logic of the mind." *Amer. Scientist* 54 (1966) 1-14.

Bruter, C.P. *Sur la nature des mathématiques.* Gauthier-Villars, Paris, France, 1973.

Cartwright, Mary L. "The mathematical mind." *Math. Spectrum* 2 (1969-70) 37-45.

Cartwright, Mary L. "Mathematics and thinking mathematically." *Amer. Math. Monthly* 77 (1970) 20-28.

Davis, Philip J. and Hersh, Reuben. *The Mathematical Experience.* Birkhäuser, Cambridge, Massachusetts, 1981.

Fisher, Charles S. "Some social characteristics of mathematicians and their work." *Amer. J. Sociology* 78 (1973) 1094-1118.

Grabiner, Judith V. "Is mathematical truth time-dependent?" *Amer. Math. Monthly* 81 (1974) 354-365.

Hadamard, Jacques. *Psychology of Invention in the Mathematical Field.* Dover, New York, 1945.

Halmos, Paul R. "Mathematics as a creative art." *Amer. Scientist* 56 (1968) 375-389.

Hahn, Hans. "Geometry and intuition." *Scientific American* 190 (April 1954) 84-91, 108; also in M. Kline. *Mathematics in the Modern World.* W.H. Freeman, San Francisco, California, 1968, pp. 184-188, 399.

Hardy, G.H. *A Mathematician's Apology.* Cambridge University Press, Cambridge, Massachusetts, 1940; 1967; excerpted in J.R. Newman. *The World of Mathematics,* V. 4, Simon and Schuster, New York, 1956, pp. 2027-2038.

Helitzer, Florence. "A conversation with three mathematicians." *University: A Princeton Quarterly* 59 (Winter 1974) 1-5, 28-30.

Henkin, Leon. "Are logic and mathematics identical?" *Science* 138 (1962) 788-794.

Hilton, Peter J. "The art of mathematics." *Univ. of Birmingham,* 1960.

Iliev, L. "Mathematics as the science of models." *Russian Math. Surveys* 27:2 (1972) 181-189.

Jones, Landon Y., Jr. "Mathematicians: They're special." *Think* 40:4 (1974) 32-35.

Kapur, J.N. *Thoughts on the Nature of Mathematics.* Atma Ram, Delhi, India, 1973.

Lefschetz, Solomon. "The structure of mathematics." *Amer. Scientist* 38 (1950) 105-111.

Newman, M.H.A. "What is mathematics? New answers to an old question." *Math. Gazette* 43 (1959) 161-171.

Otte, Michael. *Mathematiker über die Mathematik.* Springer-Verlag, New York, 1974.

Poincaré, Henri. "Mathematical creation." *Scientific American* 179 (August 1948) 54-57; also in M. Kline. *Mathematics in the Modern World.* W.H. Freeman, San Francisco, California, 1968, pp. 14-17; and in J.R. Newman, *The World of Mathematics,* V. 4, Simon and Schuster, New York, 1956, pp. 2041-2050.

Rényi, Alfred. "A Socratic dialogue on mathematics." *Canad. Math. Bull.* 7 (1964) 441-462; also in A. Rényi. *Dialogues on Mathematics.* Holden-Day, San Francisco, California, 1967, pp. 3-25.

Stein, Sherman K. "The mathematician as an explorer." *Scientific American* 204 (May 1961) 148-158, 206.

Stone, Marshall H. "The revolution in mathematics." *Liberal Education,* 47 (1961) 304-327; also in *Amer. Math. Monthly* 68 (1961) 715-734.

von Neumann, John. "The mathematician." in R.B. Heywood, *The Works of the Mind.* University of Chicago Press, Chicago, Illinois, 1947, pp. 180-196; also in J.R. Newman, *The World of Mathematics,* V. 4, Simon and Schuster, New Yok, 1956, pp. 2053-2063.

Weidman, Donald R. "Emotional perils of mathematics." *Science* 149 (1965) 1048.

Weissinger, Johannes. "The characteristic features of mathematical thought." in T.L. Saaty and F.J. Weyl, *The Spirit and Uses of the Mathematical Sciences.* McGraw-Hill, New York, 1969, pp. 9-27.

Weyl, Hermann. "The mathematical way of thinking." *Science* 92 (1940) 437-446; also in *Studies in the History of Science.* University of Pennsylvania Press, 1941, pp. 103-123.

Weyl, Hermann. "Insight and reflection." in T.L. Saaty and F.J. Weyl. *The Spirit and Uses of the Mathematical Sciences.* McGraw-Hill, New York, 1969, pp. 281-301.

Wilder, Raymond L. "The role of the axiomatic method." *Amer. Math. Monthly* 74 (1967) 115-127; also in *Math. Teaching* 41 (1967) 32-40.

Survey Papers

Bochner, Salomon. "Mathematics." *McGraw-Hill Encyclopedia of Science and Technology* 8 (1960) 175-180.

Dieudonné, Jean A. "Recent developments in mathematics." *Amer. Math. Monthly* 71 (1964) 239-248.

Eves, Howard W. "Mathematics." *Encyclopedia Americana* 18 (1976) 431-434.

Ficken, F.A. "Mathematics and the layman." *Amer. Scientist* 52 (1964) 419-430.

MacLane, Saunders. "Mathematical models of space." *Amer. Scientist* 53 (1980) 252.

Meserve, Bruce E. "New mathematics." *Encyclopedia Americana* 20 (1976) 202-205.

Meserve, Bruce E. "Number systems and notation." *Encyclopedia Americana* 20 (1976) 536f-536j.

Murray, Francis J. and Ford, Lester R. "Mathematics as a calculatory science." *Encyclopaedia Britannica,* 15th ed., 1974, Macropaedia V. 11, pp. 671-696.

Richards, Ian. "Impossibility." *Math. Magazine* 48 (1975) 249-262.

Stone, Marshall H. "The future of mathematics." *J. Math. Soc. Jap.* 9 (1957) 493-507.

Temple, G. "The growth of mathematics." *Math. Gazette* 41 (1957) 161-168.

Weil, André. "The future of mathematics." *Amer. Math. Monthly* 57 (1950) 295-306; also in F. LeLionnais. *Great Currents of Mathematical Thought,* V. 1. Dover, New York, 1971, pp. 321-336.

Advanced Exposition

Abbott, J.C. *The Chauvenet Papers: A Collection of Prize-Winning Expository Papers in Mathematics,* 2 vols. Mathematical Association of America, Washington, D.C., 1978.

Aleksandrov, A.D., Kolmogorov, A.N., Lavrent'ev, M.A. *Mathematics, Its Content, Methods, and Meaning,* 3 vols. MIT Press, Cambridge, Massachusetts, 1969.

Behnke, H., et al. *Fundamentals of Mathematics,* 3 vols. MIT Press, Cambridge, Massachusetts, 1974.

Saaty, Thomas L. *Lectures on Modern Mathematics,* 3 vols. John Wiley, New York, 1963-1965.

Biography and Autobiography

Bell, Eric Temple. *Men of Mathematics.* Simon and Schuster, New York, 1937.

Box, Joan Fisher. *R.A. Fisher, The Life of a Scientist.* Wiley, New York, 1978.

Dauben, Joseph Warren. *Georg Cantor: His Mathematics and Philosophy of the Infinite.* Harvard University Press, Cambridge, Massachusetts, 1979.

Grattan-Guinness, Ivor. *Joseph Fourier, 1768-1830.* MIT Press, Cambridge, Massachusetts, 1972.

Halmos, Paul R. "Nicolas Bourbaki." *Scientific American* 196 (May 1957) 88-99, 174.

Halmos, Paul R. "The legend of John von Neumann." *Amer. Math. Monthly* 80 (1973) 382-394.

Hardy, G.H. *Ramanujan.* Chelsea Publishing, New York, 1968.

Hoffman, Banesh. *Albert Einstein, Creator and Rebel.* Viking Press, New York, 1972.

Infeld, Leopold. *Whom the Gods Love.* Whittlesey House, 1948; NCTM, Reston, Virginia, 1978.

Kovalevskaya, Sofya. *Sofya Kovalevskaya: A Russian Childhood.* Springer-Verlag, New York, 1978.

Mahoney, Michael S. *The Mathematical Career of Pierre de Fermat.* Princeton University Press, Princeton, New Jersey, 1973.

Meschkowski, Herbert. *Ways of Thought of Great Mathematicians.* Holden-Day, San Francisco, California, 1964.

Morgan, Bryan. *Men and Discoveries in Mathematics.* Transatlantic Arts, Inc., Levittown, New York, 1972.

Morse, Philip M. *In at the Beginnings: A Physicist's Life.* MIT Press, Cambridge, Massachusetts, 1977.

Ore, Oystein. *Niels Henrik Abel, Mathematician Extraordinary.* University of Minnesota Press, Minneapolis, Minnsota, 1957; Chelsea Press, New York, 1974.

Osen, Lynn M. *Women in Mathematics.* MIT Press, Cambridge, Massachusetts, 1974.

Perl, Teri. *Math Equals: Biographies of Women Mathematicians and Related Activities.* Addison-Wesley, Reading, Massachusetts, 1978.

Reid, Constance. *Courant in Göttingen and New York.* Springer-Verlag, New York, 1976.

Reid, Constance. *Hilbert.* Springer-Verlag, New York, 1970.

Ulam, S.M. *Adventures of a Mathematician.* Charles Scribner's Sons, New York, 1976.

Wiener, Norbert. *Ex-Prodigy.* Simon and Schuster, New York, 1953; MIT Press, Cambridge, Massachusetts, 1964.

Wiener, Norbert. *I Am a Mathematician.* Doubleday, New York, 1956; MIT Press, Cambridge, Massachusetts, 1964.

History

Al-Daffa, Ali Abdullah. *The Muslim Contribution to Mathematics.* Humanities Press, Atlantic Highlands, New Jersey, 1977.

Bell, Eric Temple. *The Development of Mathematics.* McGraw-Hill, New York, 1945.

Boyer, Carl B. *A History of Mathematics.* John Wiley, New York, 1968.

Chace, Arnold Buffum. *The Rhind Mathematical Papyrus.* NCTM, Reston, Virginia, 1979.

Goldstine, H. *The Computer from Pascal to von Neumann.* Princeton University Press, Princeton, New Jersey, 1972.

Kline, Morris. *Mathematical Thought from Ancient to Modern Times.* Oxford University Press, New York, 1972.

Kline, Morris. *Mathematics in Western Culture.* Oxford University Press, New York, 1953; 1964.

Kramer, Edna E. *The Nature and Growth of Modern Mathematics.* Hawthorn, New York, 1970; Fawcett, New York, 1973.

Lambert, Joseph B., et al. "Maya arithmetic." *Amer. Scientist* 68 (May-June 1980) 249-255.

LeVeque, William J., et al. "History of mathematics." *Encyclopaedia Britannica,* 15th ed., 1974, Macropaedia, V. 11, pp. 639-670.

Menninger, Karl. *Number Words and Number Symbols, A Cultural History of Numbers.* MIT Press, Cambridge, Massachusetts, 1977.

Resnikoff, H.L. and Wells, R.O., Jr. *Mathematics in Civilization: Geometry and Calculation as Keystones of Culture.* Holt, Rinehart and Winston, New York, 1973.

Weyl, Hermann. "A half-century of mathematics." *Amer. Math. Monthly* 58 (1951) 523-553.

Wilder, Raymond L. "The origin and growth of mathematical concepts." *Bull. Amer. Math. Soc.* 59 (1963) 423-448.

Wilder, Raymond L. *Evolution of Mathematical Concepts.* Halsted Press, New York, 1974.

Wilder, Raymond L. "History in the mathematics curriculum: Its status, quality, and function." *Amer. Math. Monthly* 79 (1972) 479-495.

Zaslavsky, Claudia. *Africa Counts: Number and Pattern in African Culture.* Prindle, Weber and Schmidt, Boston, Massachusetts, 1973.

Mathematics and Science

Birkhoff, George D. "The mathematical nature of physical theories." *Amer. Scientist* 31 (1943) 281-310.

Browder, Felix E. "Is mathematics relevant? And if so, to what?" *University of Chicago Magazine* 67:3 (Spring 1975) 11-16; also appears as "The relevance of mathematics." *Amer. Math. Monthly* 83 (1976) 249-254.

Calder, Nigel. *Einstein's Universe.* Viking Press, New York, 1979.

Courant, Richard. "Mathematics in the modern world." *Scientific American* 211 (September 1964) 40-49, 269; also in M. Kline, *Mathematics in the Modern World.* W.H. Freeman, San Francisco, California, 1968, pp. 19-27, 394.

De Broglie, Louis. "The role of mathematics in the development of contemporary theoretial physics." in F. Le-Lionnais, *Great Currents on Mathematical Thought,* V. 2, Dover, New York, 1971, pp. 78-93.

Gardner, Martin. *The Ambidextrous Universe: Mirror Asymmetry and Time-Reversed Worlds,* Second Revised, Updated Edition. Scribner's, New York, 1979.

Penrose, Roger. "Einstein's vision and the mathematics of the natural world." *The Sciences* 19 (March 1979) 6-9.

Pólya, George. *Mathematical Methods in Science.* Mathematical Association of America, Washington, D.C., 1977.

Schwartz, Jacob T. "The pernicious influence of mathematics on science." in E. Nagel, P. Suppes and A. Tarski, *Logic, Methodology and Philosophy of Science.* Stanford University Press, Stanford, California, 1962, pp. 356-360.

Stone, Marshall H. "Mathematics and the future of science." *Bull. Amer. Math. Soc.* 63 (1957) 61-76.

Suppes, P. "A comparison of the meaning and uses of models in mathematics and the empirical sciences." *Synthese* 12 (1960) 287-301.

Wigner, Eugene P. "The unreasonable effectiveness of mathematics in the natural sciences." *Comm. Pure Appl. Math.* 13 (1960) 1-14; also in T.L. Saaty and F.J. Weyl. *The Spirit and Uses of the Mathematical Sciences.* McGraw-Hill, New York, 1969, pp. 123-140; in *Studies in Mathematics,* V. 16, SMSG, Stanford, California, 1967, pp. 31-44; and in E.P. Wigner, *Symmetries and Reflections: Scientific Essays of Eugene P. Wigner,* Indiana University Press, Bloomington, Indiana, 1967, pp. 222-237.

Zukav, Gary. *The Dancing Wu Li Masters: An Overview of the New Physics.* William Morrow, New York, 1979.

Mathematics and Society

Booss, Bernhelm and Niss, Mogens (eds.). *Mathematics and the Real World*. Birkhäuser, Boston, Massachusetts, 1979.

Fehr, Howard F. "Value and the study of mathematics." *Scripta Math.* 21 (1955) 49-53.

Morse, Marston. "Mathematics in our culture." in T.L. Saaty and F.J. Weyl, *The Spirit and Uses of the Mathematical Sciences*. McGraw-Hill, New York, 1969, pp. 105-120.

Whitehead, Alfred North. "Mathematics and liberal education." in A.N. Whitehead, *Essays in Science and Philosophy*. Philosophical Library, New York, 1947, pp. 175-188.

Whitehead, Alfred North. "Mathematics as an element in the history of thought." in J.R. Newman, *The World of Mathematics*, V. 1. Simon and Schuster, New York, 1956, pp. 402-416.

Wilder, Raymond L. "Trends and social implications of research." *Bull. Amer. Math. Soc.* 75 (1969) 891-906.

Philosophy and Logic

Baum, Robert J. *Philosophy and Mathematics: From Plato to the Present*. Freeman, Cooper and Company, San Francisco, California, 1974.

Benacerraf, Paul and Putnam, Hilary. *Philosophy of Mathematics: Selected Readings*. Prentice-Hall, Englewood Cliffs, New Jersey, 1964.

Crossley, J.N., et al. *What is Mathematical Logic?* Oxford University Press, New York, 1972.

Hofstadter, Douglas R. *Gödel, Escher, Bach: An Eternal Golden Braid*. Basic Books, New York, 1979.

Lakatos, Imre. "Proofs and refutations." *Brit. J. Phil. Science* 14 (1963-64) 1-25, 120-139, 221-245, 296-342. Also available as: *Proofs and Refutations: The Logic of Mathematical Discovery*. Cambridge University Press, New York, 1976.

Nagel, Ernest and Newman, James R. *Gödel's Proof*. New York University Press, New York, 1958.

Wilder, Raymond L. "The nature of mathematical proof." *Amer. Math. Monthly* 51 (1944) 309-323.

Wilder, Raymond L. *Introduction to the Foundations of Mathematics*, 2nd ed. John Wiley, New York, 1965.

Symmetry, Art, Aesthetics

Baker, Lillian F.; Schattschneider, Doris J. *The Perspective Eye: Art and Math*. Allentown Art Museum, Allentown, Pennsylvania, 1979.

Bezuszka, Stanley; Kenney, Margaret; and Silvey, Linda. *Tessellations: The Geometry of Patterns*. Creative Publications, California, 1977.

Birkhoff, George D. *Aesthetic Measure*. Harvard University Press, Cambridge, Massachusetts, 1933.

deFinetti, Bruno. *Die Kunst des Sehens in der Mathematik*. Birkhäuser, Basel, Switzerland, 1974.

Holden, Alan. *Shapes, Space, and Symmetry*. Columbia University Press, New York, 1971.

Huntley, H.E. *The Divine Proportion: A Study in Mathematical Beauty*. Dover, New York, 1970.

Lanczos, Cornelius. *Space Through the Ages*. Academic Press, New York, 1970.

Linn, Charles F. *The Golden Mean: Mathematics and the Fine Arts*. Doubleday and Company, New York, 1974.

Lockwood, E.H. and Macmillan, R.H. *Geometric Symmetry*. Cambridge University Press, New York, 1978.

Loeb, Arthur L. *Space Structures, Their Harmony and Counterpoint*. Addison-Wesley, Reading, Massachusetts, 1976.

Malina, Frank J., ed. *Visual Art, Mathematics, and Computers: Selections from the Journal Leonardo*. Pergamon, Elmsford, New York, 1979.

Mandelbrot, Benoit. *Fractals: Form, Chance, and Dimension*. W.H. Freeman, San Francisco, California, 1977.

Moineau, J.-C. *Mathématique de l'esthétique*. Dunod, Paris, France, 1969.

Ouchi, Hajime. *Japanese Optical and Geometrical Art*. Dover, New York, 1977.

Pearce, Peter. *Structure in Nature as a Strategy for Design*. MIT Press, Cambridge, Massachusetts, 1978.

Pearce, Peter and Pearce, Susan. *Polyhedra Primer*. D. Van Nostrand, New York, 1978.

Pedoe, Dan. *Geometry and the Liberal Arts*. St. Martin's Press, New York, 1978.

Pugh, Anthony. *Polyhedra, A Visual Approach*. University of California Press, California, 1976.

Robson, Ernest and Wimp, Jet, eds. *Against Infinity: An Anthology of Contemporary Mathematical Poetry*. Primary Press, Pennsylvania, 1979.

Rosen, Joe. *Symmetry Discovered: Concepts and Applications in Nature and Science*. Cambridge University Press, New York, 1975.

Senechal, Marjorie and Fleck, George. *Patterns of Symmetry*. University of Massachusetts Press, Amherst, Massachusetts, 1977.

Shubnikov, A.V. and Koptsik, V.A. *Symmetry in Science and Art*. Plenum Press, New York, 1974.

Stevens, Peter S. *Patterns in Nature*. Little, Brown and Company, Boston, Massachusetts, 1974.

Wechsler, Judith, ed. *On Aesthetics in Science*. MIT Press, Massachusetts, 1978.

Weyl, Hermann. *Symmetry*. Princeton University Press, Princeton, New Jersey, 1952; excerpted in J.R. Newman. *The World of Mathematics*, V. 1. Simon and Schuster, New York, 1956, pp. 671-724.

Computing

Feldman, Jerome A. "Programming languages." *Scientific American* 241 (December 1979) 94-116.

Hamming, Richard W. "Intellectual implications of the computer revolution." *Amer. Math. Monthly* 70 (1963) 4-11; also in T.L. Saaty and F.J. Weyl. *The Spirit and Uses of the Mathematical Sciences*, McGraw-Hill, 1969, pp. 188-199; in *Studies in Mathematics*, V. 16, School Mathematics Study Group, Stanford, California, 1967, pp. 45-52; and in Z.W. Pylyshyn. *Perspectives on the Computer Revolution.* Prentice-Hall, Englewood Cliffs, New Jersey, 1970, pp. 370-377.

McCorduck, Pamela. *Machines Who Think: A Personal Inquiry into the History of Prospects of Artificial Intelligence.* Freeman, San Francisco, California, 1979.

Weizenbaum, Joseph. *Computer Power and Human Reason: From Judgment to Calculation.* W.H. Freeman, San Francisco, California, 1976.

Pedagogy

Dieudonné, Jean A. "Should we teach modern mathematics?" *Amer. Scientist* 61 (1973) 16-19.

Engel, Arthur. "The relevance of modern fields of applied mathematics for mathematical education." *Educ. Studies Math.* 2 (1969-70) 257-269.

Henrici, Peter. "Reflections of a teacher of applied mathematics." *Quarterly of Applied Math.* 30 (1972) 31-39.

Hilton, Peter J. "The survival of education." *Educ. Tech.* 13:11 (November 1973) 12-16.

Kemeny, John G. "Teaching the new mathematics." *Atlantic Monthly* 210 (October 1962) 90-91+; also in J.G. Kemeny. *Random Essays on Mathematics, Education and Computers.* Prentice-Hall, Englewood Cliffs, New Jersey, 1964, pp. 27-34.

Klamkin, Murray S. "The teaching of mathematics so as to be useful." *Educ. Studies Math.* 1 (1968-69) 126-160.

Kline, Morris. *Why the Professor Can't Teach: Mathematics and the Dilemma of University Education.* St. Martin's Press, New York, 1977.

Kline, Morris. "Logic versus pedagogy." *Amer. Math. Monthly* 77 (1970) 264-282.

Lazarus, Mitchell. "Mathophobia: Some personal speculations." *Nat. Elem. Principal* 53:2 (Jan.-Feb. 1974) 16-22.

Lighthill, M.J. "The art of teaching the art of applying mathematics." *Math. Gazette* 55 (1971) 249-270.

Ordman, Edward T. "One and one is nothing: Liberating mathematics." *Soundings* 56 (1973) 164-181.

Pollak, Henry O. "How can we teach applications of math?" *Educ. Studies Math.* 2 (1969-70) 393-404.

Pollak, Henry O. "On some of the problems of teaching applications of mathematics." *Educ. Studies Math.* 1 (1968-69) 24-30.

Pólya, George. *How to Solve It.* Princeton University Press, Princeton, New Jersey, 1945; excerpted in J.R. Newman. *The World of Mathematics,* V. 3. Simon and Schuster, New York, 1956, pp. 1980-1992.

Pólya, George. *Mathematical Discovery,* 2 vols. John Wiley, New York, 1962 and 1965.

Pólya, George. *Mathematics and Plausible Reasoning,* Vols. I and II. Princeton University Press, Princeton, New Jersey. Vol. I, 1954; Vol. II, rev. ed., 1969.

Puzzles & Recreations

Ball, W.W. Rouse and Coxeter, H.S. MacDonald. *Mathematical Recreations and Essays,* Twelfth Edition. University of Toronto Press, Toronto, Canada, 1974.

Dudeney, Henry Ernest. *The Canterbury Puzzles and Other Curious Problems,* Fourth Edition. Dover, New York, 1958.

Duffin, R.J. *Puzzles, Games, and Paradoxes.* Carnegie-Mellon University, Pittsburgh, Pennsylvania, 1979.

Fixx, James E. *Solve It! A Perplexing Profusion of Puzzles.* Doubleday, New York, 1978.

Fujimura, Kobon. *The Tokyo Puzzles.* Charles Scribner's Sons, New York, 1978.

Gardner, Martin. *Martin Gardner's Sixth Book of Mathematical Games from Scientific American.* W.H. Freeman, San Francisco, California, 1971.

Gardner, Martin. *Mathematical Carnival.* Alfred A. Knopf, Inc., New York, 1975.

Gardner, Martin. *Mathematical Circus.* Alfred A. Knopf, Inc., New York, 1979.

Gardner, Martin. *Mathematical Magic Show.* Alfred A. Knopf, New York, 1977.

Gardner, Martin. *Mathematics, Magic and Mystery.* Dover, New York, 1956.

Gardner, Martin. *New Mathematical Diversions from Scientific American.* Simon and Schuster, New York, 1966.

Gardner, Martin. *The Numerology of Dr. Matrix.* Simon and Schuster, New York, 1967.

Gardner, Martin. *The Scientific American Book of Mathematical Puzzles and Diversions.* Simon and Schuster, New York, 1959.

Gardner, Martin. *The Second Scientific American Book of Mathematical Puzzles and Diversions.* Simon and Schuster, New York, 1961.

Gardner, Martin. *The Unexpected Hanging, and Other Mathematical Diversions.* Simon and Schuster, New York, 1969.

Hunter, J.A.H. *Mathematical Brain-Teasers.* Dover, New York, 1976.

Hunter, J.A.H. and Madachy, Joseph S. *Mathematical Diversions.* Dover, New York, 1975.

Kraitchik, Maurice. *Mathematical Recreations,* 2nd ed. Dover, New York, 1953.

Mott-Smith, Geoffrey. *Mathematical Puzzles for Beginners and Enthusiasts,* Second Revised Edition. Dover, New York, 1954.

Ogilvy, C. Stanley. *Tomorrow's Math: Unsolved Problems for the Amateur,* Second Edition. Oxford University Press, New York, 1972.

Schwartz, Benjamin L., ed. *Mathematical Solitaires & Games.* Baywood Publishers, Farmingdale, New York, 1980.

Singmaster, David. *Notes on Rubik's Magic Cube,* Fifth Edition, Preliminary Version. Polytechnic of the South Bank, London, 1980.

Smullyan, Raymond. *The Chess Mysteries of Sherlock Holmes.* Alfred A. Knopf, New York, 1979.

Smullyan, Raymond M. *What is the Name of This Book?* Prentice-Hall, Englewood Cliffs, New Jersey, 1978.

Tietze, Heinrich. *Famous Problems of Mathematics.* Graylock Press, Baltimore, Maryland, 1965.

Reference

A Basic Library List for Four-Year Colleges, Second Edition. Mathematical Association of America, Washington, D.C., 1976.

A Basic Library List for Two-Year Colleges, Second Edition. Mathematical Association of America, Washington, D.C., 1980.

Gaffney, Matthew P. and Steen, Lynn Arthur. *Annotated Bibliography of Expository Writing in the Mathematical Sciences.* Mathematical Association of America, Washington, D.C., 1976.

Hoyrup, Else. *Books About Mathematics: History, Philosophy, Education, Models, System Theory, and Works of Reference, etc.: A Bibliography.* Roskilde University Center, Denmark, 1979.

Hoyrup, Else. *Women and Mathematics, Science and Engineering.* Roskilde University Center, Denmark, 1978.

May, Kenneth O. *Bibliography and Research Manual of the History of Mathematics.* University of Toronto Press, Toronto, Canada, 1973.

May, Kenneth O. *Index of the American Mathematical Monthly, V. 1-80. (1894-1973).* Mathematical Association of America, Washington, D.C., 1977.

Schaaf, William L. *A Bibliography of Recreational Mathematics,* V. 1-4. National Council of Teachers of Mathematics, Reston, Virginia, 1954, 1970, 1973, 1978.

Schaaf, William L. *Mathematics and Science: An Adventure in Postage Stamps.* National Council of Teachers of Mathematics, Reston, Virginia, 1978.

Schaefer, Barbara Kirsch. *Using the Mathematical Literature, A Practical Guide.* Dekker, New York, 1979.

Schneider, David. *Annotated Bibliography of Films and Videotapes for College Mathematics.* Mathematical Association of America, Washington, D.C., 1980.

Seebach, J. Arthur and Steen, Lynn Arthur. *Mathematics Magazine: 50 Year Index (1926-1977).* Mathematical Association of America, Washington, D. C., 1979.

Singmaster, David. *List of 16mm Films on Mathematical Subjects.* Open University, England.

Fiction, Fables and Anecdotes

Abbott, Edwin A. *Flatland–A Romance of Many Dimensions.* Little, Brown, Boston, Massachusetts, 1928; Dover, New York, 1952.

Eves, Howard W. *In Mathematical Circles.* Prindle, Weber and Schmidt, Boston, Massachusetts, 1969.

Eves, Howard W. *Mathematical Circles Revisited.* Prindle, Weber and Schmidt, Boston, Massachusetts, 1971.

Eves, Howard W. *Mathematical Circles Squared.* Prindle, Weber and Schmidt, Boston, Massachusetts, 1971.

Eves, Howard W. *Mathematical Circles Adieu.* Prindle, Weber and Schmidt, Boston, Massachusetts, 1977.

Fadiman, Clifton. *Fantasia Mathematica.* Simon and Schuster, New York, 1961.

Fadiman, Clifton. *The Mathematical Magpie.* Simon and Schuster, New York, 1962.

Moritz, Robert Edouard. *On Mathematics: A Collection of Witty, Profound, Amusing Passages About Mathematics and Mathematicians.* Dover, New York, 1942.

Probability and Statistics

David, F.N. *Games, Gods and Gambling.* Hafner Press, New York, 1962.

Huff, Darrel and Geis, Irving. *How to Lie with Statistics.* W.W. Norton, New York, 1954.

Kimble, Gregory A. *How to Use (and Misuse) Statistics.* Prentice-Hall, New Jersey, 1978.

Levinson, Horace C. *Chance, Luck, and Statistics,* Second Edition. Dover, New York, 1963.

Moore, David S. *Statistics: Concepts and Controversies.* Freeman, San Francisco, California, 1979.

Tanur, Judith M., ed. *Statistics: A Guide to the Unknown,* Second Edition. Holden-Day, California, 1978.

Williams, Bill. *A Sampler on Sampling.* Wiley, New York, 1978.

Topology and Geometry

Barr, Stephen. *Experiments in Topology.* Thomas Y. Crowell, New York, 1972.

Engel, Kenneth. "Shadows of the 4th dimension." *Science 80* (July/August 1980) 68-73.

Flegg, H. Graham. *From Geometry to Topology.* English University Press (Crane Rusak, New York, distributor), 1974.

Gray, Jeremy. *Ideas of Space: Euclidean, Non-Euclidean, and Relativistic.* Clarendon Press, New York, 1979.

Griffiths, H.B. *Surfaces.* Cambridge University Press, New York, 1976.

Pedoe, D. *Circles, A Mathematical View.* Dover, New York, 1979.

Wenninger, Magnus J. *Polyhedron Models.* Cambridge University Press, New York, 1971.

Wenninger, Magnus J. *Spherical Models.* Cambridge University Press, New York, 1979.

Miscellaneous Books

Asimov, Isaac. *Asimov on Numbers.* Doubleday and Company, New York, 1977.

Brams, Steven J. *Spatial Models of Election Competition.* EDC/UMAP, Newton, Massachusetts, 1979.

Brams, Steven J. *Biblical Games: A Strategic Analysis of Stories in the Old Testament.* MIT Press, Cambridge, Massachusetts, 1980.

Beck, Anatole, Bleicher, Michael N. and Crowe, Donald W. *Excursions Into Mathematics.* Worth Publishers, New York, 1969.

De Morgan, Augustus. *A Budget of Paradoxes.* Open Court, LaSalle, Illinois, 1872; 1915; excerpted in J.R. Newman. *The World of Mathematics,* V. 4. Simon and Schuster, New York, 1956, pp. 2369-2382.

Honsberger, Ross. *Mathematical Gems from Elementary Combinatorics, Number Theory, and Geometry.* Mathematical Association of America, Washington, D.C., 1973.

Honsberger, Ross. *Ingenuity in Mathematics.* Mathematical Association of America, Washington, D.C., 1975.

Honsberger, Ross. *Mathematical Gems II.* Mathematical Association of America, Washington, D.C., 1976.

Honsberger, Ross. *Mathematical Morsels.* Mathematical Association of America, Washington, D.C., 1979.

Honsberger, Ross, ed. *Mathematical Plums.* Mathematical Association of America, Washington, D.C., 1979.

Kac, Mark and Ulam, Stanislaw M. *Mathematics and Logic: Retrospect and Prospects.* Frederick A. Praeger, New York, 1969.

Knuth, D.E. *Surreal Numbers.* Addison-Wesley, Reading, Massachusetts, 1974.

Kogelman, Stanley; Warren, Joseph. *Mind over Math.* Dial Press, New York, 1978. See also: Hilton, Peter and Pedersen, Jean. "Review of 'Overcoming Math Anxiety' and 'Mind over Math'." *Amer. Math. Monthly* 87 (1980) 143-148.

Lieber, Lillian R. *Human Values and Science, Art and Mathematics.* W.W. Norton and Company, New York, 1961.

Lieber, Lillian R. *Mits, Wits, and Logic.* W.W. Norton, New York, 1947; 1954; 1960.

Lieber, Lillian R. *Take a Number.* Ronald Press, New York, 1946.

Lieber, Lillian R. *The Education of T.C. Mits.* W.W. Norton, New York, 1942.

Melzak, Z.A. *Companion to Concrete Mathematics,* 2 vols. Wiley, New York, 1973, 1976.

Paulos, John Allen. *Mathematics and Humor.* University of Chicago Press, Chicago, Illinois, 1980.

Roberts, Fred S. *Discrete Mathematical Models with Applications to Social, Biological, and Environmental Problems.* Prentice-Hall, Englewood Cliffs, New Jersey, 1976.

Rubinstein, Moshe F. *Patterns of Problem Solving.* Prentice-Hall, Englewood Cliffs, New Jersey, 1975.

Tobias, Sheila. *Overcoming Math Anxiety.* Norton, New York, 1978. See also: Hilton, Peter and Pedersen, Jean. "Review of 'Overcoming Math Anxiety' and 'Mind over Math'." *Amer. Math. Monthly* 87 (1980) 143-148.

Waisman, Friedrich. *Introduction to Mathematical Thinking: The Formation of Concepts in Modern Mathematics.* Harper & Brothers, New York, 1959.

Woodcock, Alexander and Davis, Monte. *Catastrophe Theory.* E.P. Dutton and Company, New York, 1978.